Buckle Down™

Mathematics

Level 2

2nd Edition

This book belongs to: ____________________

Helping your schoolhouse meet the standards of the statehouse™

ISBN-10: 0-7836-5779-X
ISBN-13: 978-0-7836-5779-0

2BDUS02MM01 11 12 13 14 15 16 17 18 19 20

Senior Editor: Paul Meyers; Project Editor: Lynn Tauro; Production Editor: Jennifer Rapp; Cover Design: Christina Nantz; Cover Graphic Designer: Christina Kroemer; Production Director: Jennifer Booth; Art Director: Chris Wolf; Graphic Designers: Spike Schabacker, Luke Gordon; Composition: Wyndham Books.

TABLE OF CONTENTS

To the Teacher:

Objective codes are listed for each lesson in the table of contents. The numbers in the shaded gray bar that runs across the tops of the pages in the workbook show the Objectives for a given page (see example to the right). The example below shows what each part of the code stands for.

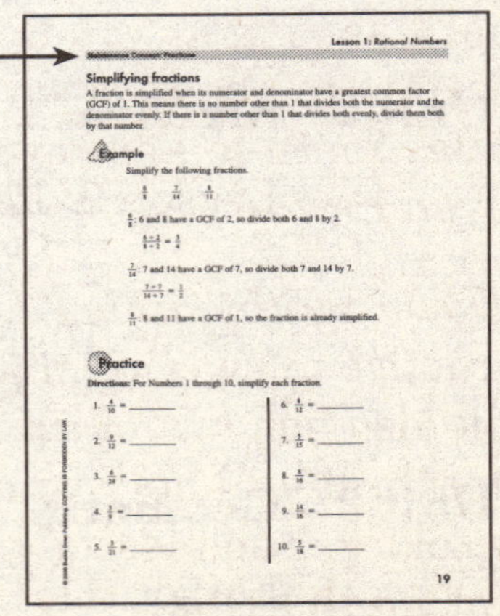

Sample code: N.B.2

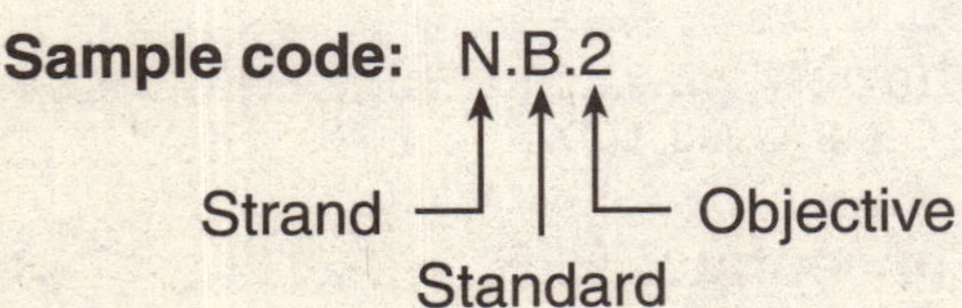

Strands:
N = Number and Operations
A = Algebra
G = Geometry
M = Measurement
D = Data Analysis and Probability

Introduction

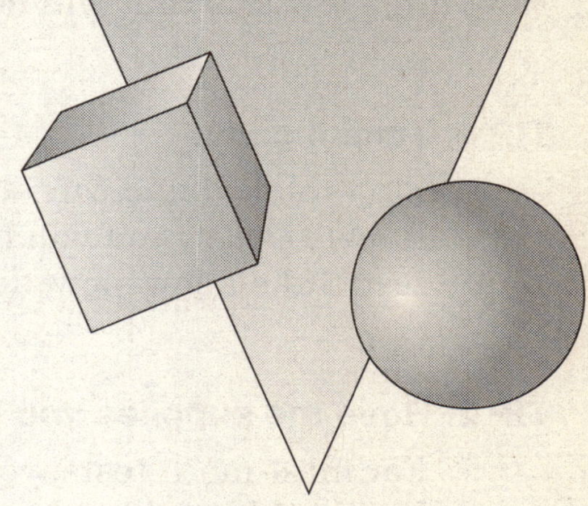

Why do we learn math? For some of us, it's just fun! For others, it can sometimes be a lot of work. But we all need to learn math. Every day, people around the world use math in many different ways.

How much money do I need to buy this milk? How tall should this building be? How can we send a rocket into space? All of these questions are answered using math. You have been learning about math as long as you have been in school, and now you are about to learn more. Someday you will use math to answer many questions—maybe even how to send a rocket into space!

Buckle Down Mathematics, Level 2, will give you lots of practice answering math questions. This workbook also has lots of tips to help you do your best at math. It will help you become a better problem solver. Each lesson will teach you important math skills.

Test-Taking Tips

Here are a few tips that will help you on test day.

TIP 1: Take it easy.

Stay relaxed and confident. Because you've practiced the problems in *Buckle Down*, you will be ready to do your best on almost any math test. Take a few slow, deep breaths before you begin the test.

TIP 2: Have the supplies you need.

For most math tests, you will need two sharp pencils and an eraser. Your teacher will tell you whether you need anything else.

TIP 3: Read the questions more than once.

Every question is different. Some questions are more difficult than others. If you need to, read a question more than once. This will help you make a plan for solving the question.

TIP 4: Learn to "plug in" answers to multiple-choice items.

When do you "plug in"? You should "plug in" whenever your answer is different from all of the answer choices or you can't come up with an answer. Plug each answer choice into the problem and find the one that makes sense. (You can also think of this as "working backwards.")

TIP 5: Check your work.

Take the time to check your work on every problem. By checking your work, you can eliminate careless mistakes.

TIP 6: Use all the test time.

Work on the test until you are told to stop. If you finish early, go back through the test and double-check your answers. You just might increase your score on the test by finding and fixing any errors you might have made.

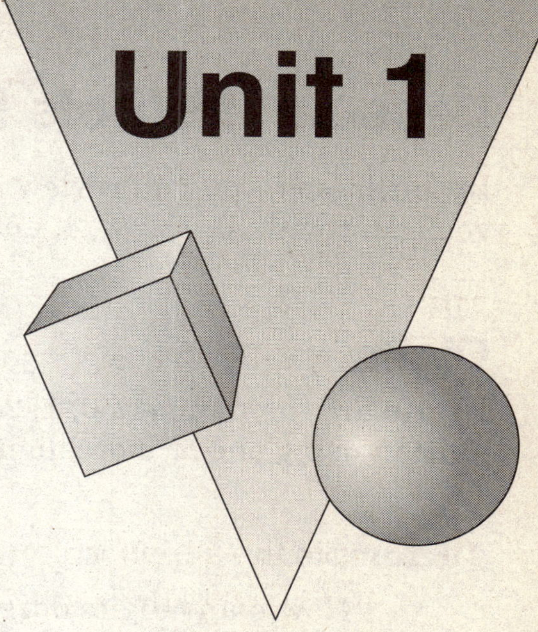

Number and Operations

Every number has a name. You know some of them. The name for 1 is "one." But what is the name for 987?

In this unit, you will show, read, and write whole numbers up to 9,999 using digits, models, and words. You will order and compare whole numbers. You will add, subtract, multiply, and divide whole numbers. You will review steps that can be used to solve word problems. You will write fractions to name a part of a whole object or a part of a set of objects. You will also count money using coins and dollar bills.

In This Unit

Lesson 1: Whole Numbers

In this lesson, you will review whole numbers. **Whole numbers** are numbers you count with: 0, 1, 2, 3, 4, and so on.

Digits

Digits are the numbers 0, 1, 2, 3, 4, 5, 6, 7, 8, and 9. Each whole number is written using one or more digits.

Example

How many digits does 138 have?

The number 138 has three digits. The digits are 1, 3, and 8.

Models

Whole numbers can be shown with **models**. You can use blocks to show models of whole numbers.

Example

This shows **one (1)**	This shows **ten (10)**	This shows **one hundred (100)**	This shows **one thousand (1,000)**

Objectives: N.A.1, N.A.2

Example

How is 138 shown in a model?

The following model shows 138.

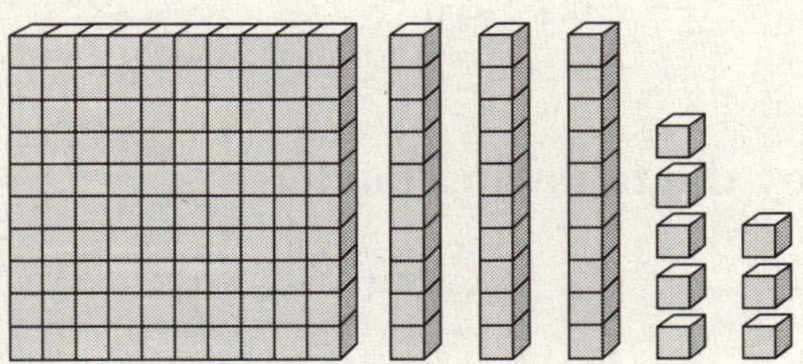

Number models make it easy to see how to write a number in **expanded form**. The number 138 written in expanded form is 100 + 30 + 8.

Words

The **word form** of a whole number is the same as how you say the number out loud. The following table shows the word form for ones, tens, hundreds, and thousands.

1	one	10	ten	100	one hundred	1,000	one thousand
2	two	20	twenty	200	two hundred	2,000	two thousand
3	three	30	thirty	300	three hundred	3,000	three thousand
4	four	40	forty	400	four hundred	4,000	four thousand
5	five	50	fifty	500	five hundred	5,000	five thousand
6	six	60	sixty	600	six hundred	6,000	six thousand
7	seven	70	seventy	700	seven hundred	7,000	seven thousand
8	eight	80	eighty	800	eight hundred	8,000	eight thousand
9	nine	90	ninety	900	nine hundred	9,000	nine thousand

Example

What is the word form of 138?

The word form of 138 is one hundred thirty-eight.

Example

What number is shown by the following model?

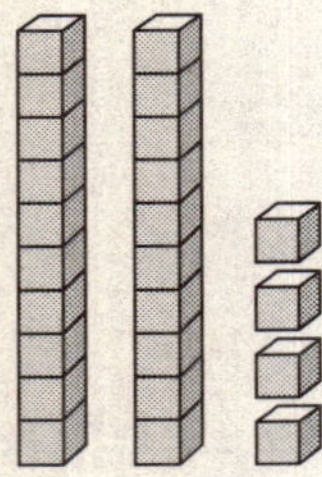

Write the **number:** 24

Write the **expanded form:** 20 + 4

Write the **word form:** twenty-four

Example

What number is shown by the following model?

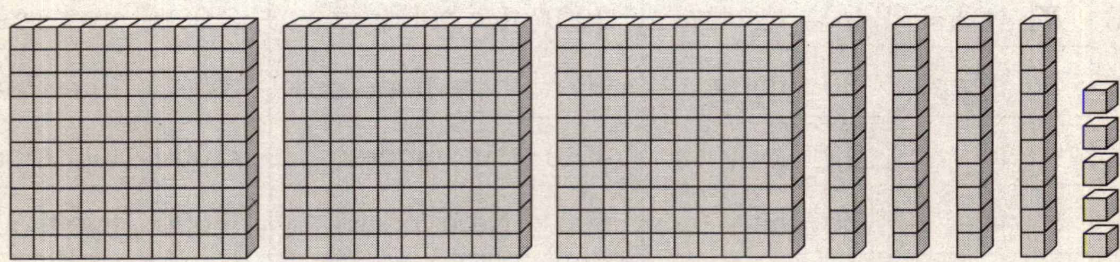

Write the **number:** 345

Write the **expanded form:** 300 + 40 + 5

Write the **word form:** three hundred forty-five

Objectives: N.A.1, N.A.2

Practice

1. What year were you born?

   Write the number. ______________________

   Write the expanded form. ______________________

   Write the word form. ______________________

2. What number is shown by the following model?

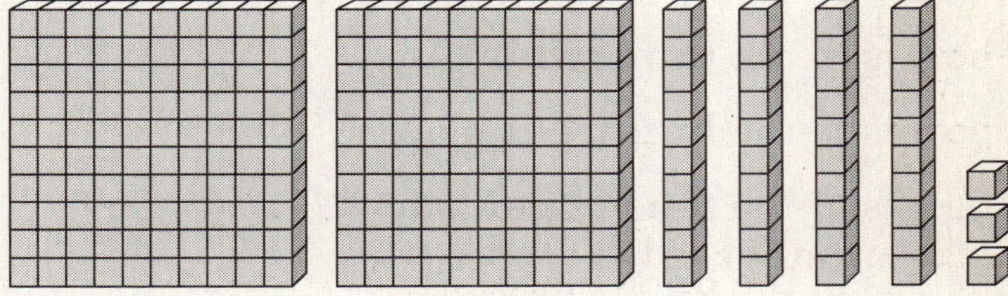

   Write the number. ______________________

   Write the expanded form. ______________________

   Write the word form. ______________________

3. What number is shown by the following model?

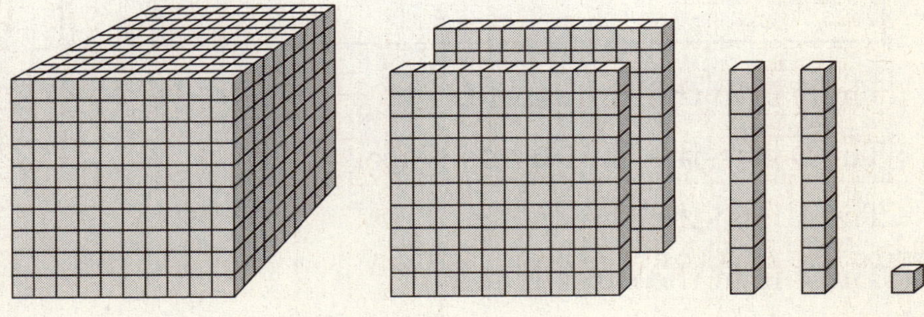

   Write the number. ______________________

   Write the expanded form. ______________________

   Write the word form. ______________________

Place Value

Each digit of a number has a **place value**. A digit's place value is where it is located in the number.

Example

The following place-value table shows the place value of each digit of 9,423.

Thousands	Hundreds	Tens	Ones
9	4	2	3

The 9 is in the **thousands** place.

The 4 is in the **hundreds** place.

The 2 is in the **tens** place.

The 3 is in the **ones** place.

Write the **word form:** nine thousand, four hundred twenty-three.

Example

The following place-value table shows the place value of each digit of 5,214.

Thousands	Hundreds	Tens	Ones
5	2	1	4

The 5 is in the **thousands** place.

The 2 is in the **hundreds** place.

The 1 is in the **tens** place.

The 4 is in the **ones** place.

Write the **word form:** five thousand, two hundred fourteen.

Objectives: N.A.1, N.A.2

Practice

Directions: Use the following place-value table to answer Numbers 1 through 9.

Thousands	Hundreds	Tens	Ones

1. Write the number eight hundred ninety in the place-value table.

2. What is the place value of the zero in eight hundred ninety?

   ____

3. What is the place value of the eight in eight hundred ninety?

   ____

4. Write the number seven hundred four in the place-value table.

5. What is the place value of the seven in seven hundred four?

   ____

6. What is the place value of the four in seven hundred four?

   ____

7. Write the number 2,643 in the place-value table.

8. What is the place value of the 4 in 2,643? ____________________

9. What is the place value of the 6 in 2,643? ____________________

10, 100, and 1,000

Numbers can be shown in different ways. Look at the models for 1, 10, 100, and 1,000 on page 4.

The model for 10 is made up of 10 of the models for 1. You can trade 1 ten for 10 ones.

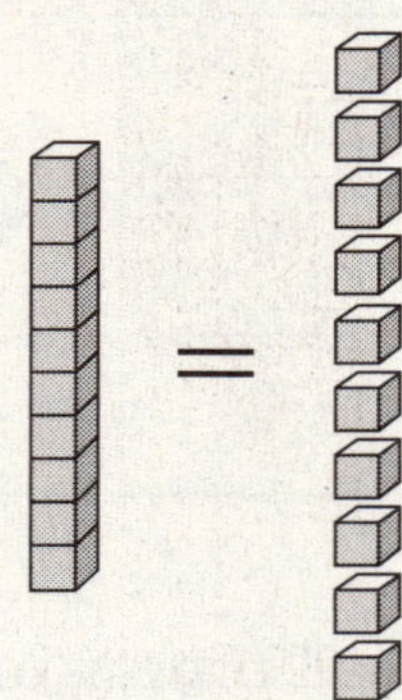

The model for 100 is made up of 10 of the models for 10. The number 100 can be thought of as 1 hundred or 10 tens. You can trade 1 hundred for 10 tens. If you trade each of the tens for 10 ones, you can see that 1 hundred is the same as 100 ones.

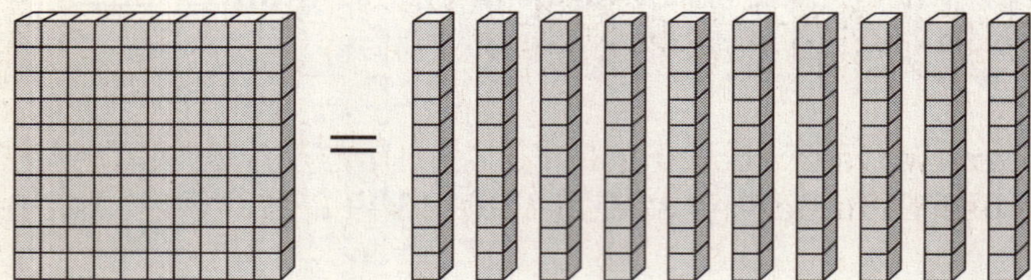

The model for 1,000 is made up of 10 of the models for 100. You can think of the number 1,000 as 1 thousand, 10 hundreds, 100 tens, or 1,000 ones. These are all the same amount.

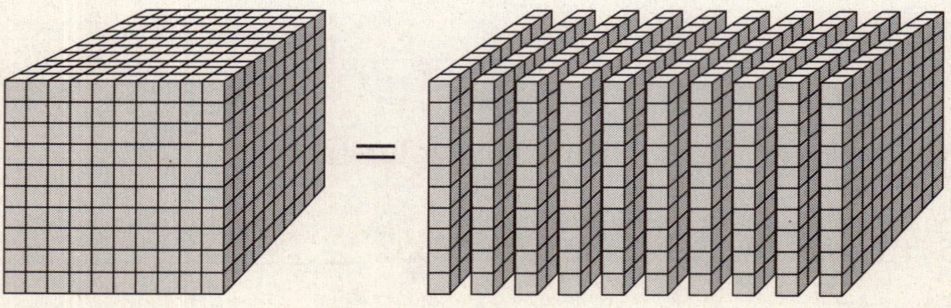

Objectives: N.A.1, N.A.2

Example

Jake and Sarah were each asked to show 1,110.

Jake showed 1,110 using 1 thousand, 1 hundred, and 1 ten.

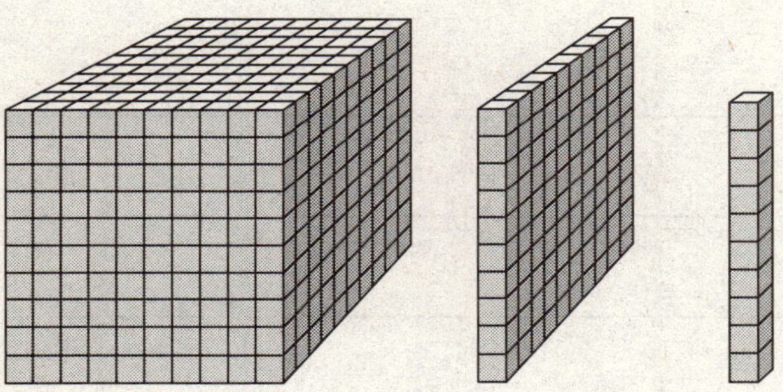

Sarah showed 1,110 using 11 hundreds and 1 ten.

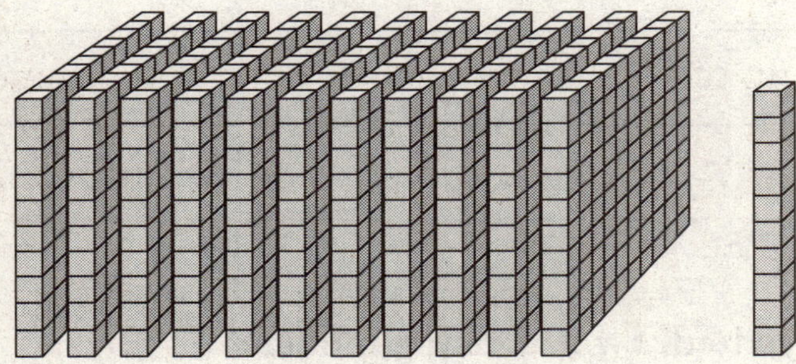

Both are ways of showing 1,110. Another way of showing 1,110 would be to use 10 hundreds and 11 tens.

Practice

Directions: For Numbers 1 and 2, write the word form of the given number. Then write two different ways of showing the number.

1. 324

   word form ______________________________

   way 1 ______________________________

   way 2 ______________________________

2. 2,010

   word form ______________________________

   way 1 ______________________________

   way 2 ______________________________

3. Show two different ways to draw 232 using models.

Objectives: N.A.5

Even and Odd Numbers

An **even** number has a 0, 2, 4, 6, or 8 in its ones place. An **odd** number has a 1, 3, 5, 7, or 9 in its ones place.

To show whether there is an even or odd number of objects, you can group the objects by 2s. If 0 objects are left over, there is an even number of objects. If 1 object is left over, there is an odd number of objects.

Example

Harold has 11 pencils. He wants to show whether he has an even or odd number of pencils. The following drawing shows how Harold grouped the pencils in groups of 2.

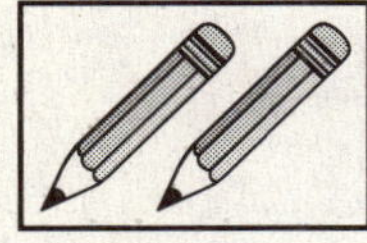  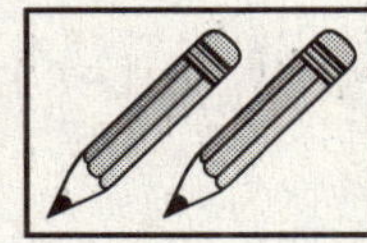 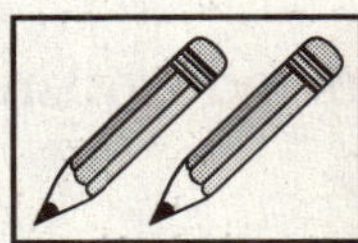 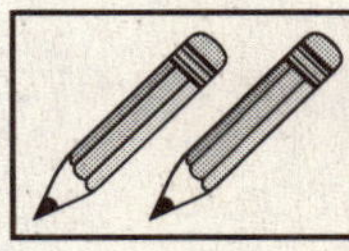 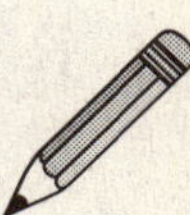

There is 1 pencil left over. Harold has shown he has an odd number of pencils.

Example

Samantha has 12 T-shirts. She wants to show whether she has an even or odd number of T-shirts. The following drawing shows how Samantha grouped her T-shirts in groups of 2.

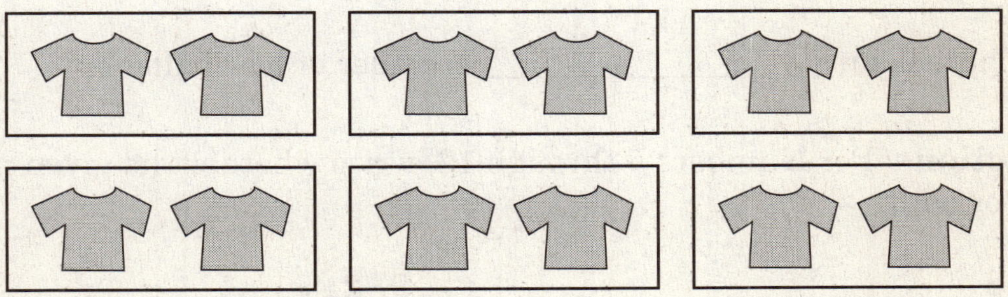

There are 0 T-shirts left over. Samantha has shown she has an even number of T-shirts.

Practice

Directions: Use the following water bottles to answer Numbers 1 through 3.

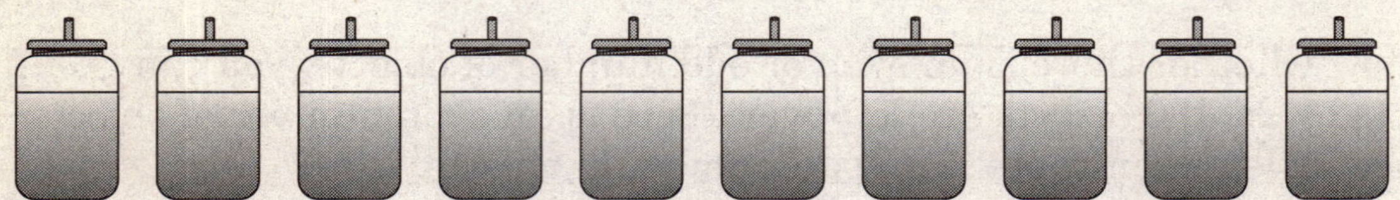

1. Group the water bottles by 2s. Draw a circle around each group.

2. How many water bottles are left over? __________

3. Is there an even or odd number of water bottles? __________

4. Is there an even or odd number of baseballs shown below?

There is an ____________________ number of baseballs.

Directions: For Numbers 5 through 12, write whether the given number is even or odd.

5. 18 ____________________

6. 67 ____________________

7. 91 ____________________

8. 170 ____________________

9. 245 ____________________

10. 556 ____________________

11. 803 ____________________

12. 932 ____________________

Objectives: N.A.4

Counting by 2s

When you count by 2s, you say only even numbers. They are shaded in the chart.

1	2	3	4	5	6	7	8	9	10
11	12	13	14	15	16	17	18	19	20
21	22	23	24	25	26	27	28	29	30
31	32	33	34	35	36	37	38	39	40
41	42	43	44	45	46	47	48	49	50
51	52	53	54	55	56	57	58	59	60
61	62	63	64	65	66	67	68	69	70
71	72	73	74	75	76	77	78	79	80
81	82	83	84	85	86	87	88	89	90
91	92	93	94	95	96	97	98	99	100

Example

You can count by 2s to find the number of seashells.

2 **4** **6** **8** **10**

There are 10 seashells.

Practice

Directions: Use the stars to answer Numbers 1 and 2.

1. Divide the stars into groups of 2. Draw a circle around each group.

2. Count by 2s to find how many stars there are altogether.

____

3. Rocco found some birds' nests in his yard. Count by 2s to find how many eggs there are altogether.

____

Directions: For Numbers 4 through 8, count by 2s to fill in the missing numbers.

4. 34, 36, ________, 40, ________, ________, 46

5. 98, ________, ________, 104, 106, ________

6. 444, 446, ________, ________, 452, ________

7. 600, ________, ________, ________, 608, 610

8. ________, 992, ________, 996, 998, ________

Objectives: N.A.4

Counting by 5s and 10s

When you count by 5s, all the numbers will end in 0 or 5. These numbers are shaded on the chart.

1	2	3	4	5	6	7	8	9	10
11	12	13	14	15	16	17	18	19	20
21	22	23	24	25	26	27	28	29	30
31	32	33	34	35	36	37	38	39	40
41	42	43	44	45	46	47	48	49	50
51	52	53	54	55	56	57	58	59	60
61	62	63	64	65	66	67	68	69	70
71	72	73	74	75	76	77	78	79	80
81	82	83	84	85	86	87	88	89	90
91	92	93	94	95	96	97	98	99	100

When you count by 10s, all the numbers will end in 0. Look at the very last column on the right side of the chart.

1	2	3	4	5	6	7	8	9	10
11	12	13	14	15	16	17	18	19	20
21	22	23	24	25	26	27	28	29	30
31	32	33	34	35	36	37	38	39	40
41	42	43	44	45	46	47	48	49	50
51	52	53	54	55	56	57	58	59	60
61	62	63	64	65	66	67	68	69	70
71	72	73	74	75	76	77	78	79	80
81	82	83	84	85	86	87	88	89	90
91	92	93	94	95	96	97	98	99	100

Practice

Directions: Use the charts on the previous page to answer Numbers 1 through 5.

1. What are you counting by if you count 15, 20, 25, 30, 35, 40? __________

2. If you count by 10s, what is the next number? 20, 30, 40, __________

3. Can you get to the number 50 if you count by 5s? __________

4. Can you get to the number 50 if you count by 10s? __________

5. You can get to some of the same numbers if you count by 5s and 10s. What are those numbers?

   ____

Directions: For Numbers 6 through 10, count by 5s or 10s to fill in the missing numbers.

6. 65, 70, ________, 80, ________, ________, 95

7. 180, ________, ________, 210, 220, ________

8. 350, 360, ________, ________, 390, ________

9. 600, ________, ________, ________, 620, 625

10. ________, 920, ________, 940, 950, ________

Objectives: N.A.4

Counting by 25s

When you need to count to a large number, such as 1,000, you can get there faster if you count by 25s. Here is a chart that shows counting by 25s.

25	50	75	100
125	150	175	200
225	250	275	300
325	350	375	400
425	450	475	500
525	550	575	600
625	650	675	700
725	750	775	800
825	850	875	900
925	950	975	1000

Practice

Directions: Use the chart above to answer Numbers 1 through 5.

1. Fill in the missing numbers. 25, 50, ________, 100, ________, ________

2. What number comes **before** 125? ________

3. What number comes **after** 800? ________

4. What numbers are **greater than** 900? ______________________________

____

5. What numbers are **less than** 150? ______________________________

____

Ordinal Numbers

Ordinal numbers tell the position or order of objects. Usually the position is named from left to right. Ordinal numbers can be written using words or numbers. The first twenty ordinal numbers are shown in the chart below.

1st	first	11th	eleventh
2nd	second	12th	twelfth
3rd	third	13th	thirteenth
4th	fourth	14th	fourteenth
5th	fifth	15th	fifteenth
6th	sixth	16th	sixteenth
7th	seventh	17th	seventeenth
8th	eighth	18th	eighteenth
9th	ninth	19th	nineteenth
10th	tenth	20th	twentieth

Example

What ordinal number describes the position of the mouse?

Begin on the left. Count over to the mouse: 1, 2, 3.

The mouse is the third (3rd) animal.

Example

Greg is standing in a long line. There are 32 people in front of him. What ordinal number describes Greg's position in line?

Since there are 32 people in front of Greg, he is number 33 in line. He is the thirty-third (33rd) person in line.

Objectives: N.A.6

Practice

Directions: Use the shapes below to answer Numbers 1 through 4.

△ ○ □ ♡ ◇ ⌓ ▭ ☆

1. What position is the ◇ in? _______________

2. What shape is in the fourth position? _______________

3. What position is the ⌓ in? _______________

4. What shape is in the eighth position? _______________

5. Kelly lined up her toys in the following order.

Write the toy that goes with each ordinal number.

1st _______________ 5th _______________ 2nd _______________

6. Justin ran a race last Saturday. He was the 54th person to cross the finish line. How many people finished in front of Justin?

____

7. Write the following ordinal numbers using words.

12th ____

40th ____

92nd ____

Mathematics Practice

1. How do you write the number one hundred fifty-three using digits?

   Ⓐ 103
   Ⓑ 115
   Ⓒ 135
   Ⓓ 153

2. Which number has a 2 in the tens place?

   Ⓐ 2,467
   Ⓑ 532
   Ⓒ 274
   Ⓓ 129

3. Each package of stickers has 25 gold stars in it. Mrs. Mason bought 5 packages of stickers.

   How many gold stars did Mrs. Mason buy?

   Ⓐ 50
   Ⓑ 100
   Ⓒ 125
   Ⓓ 150

4. What number does the following model show?

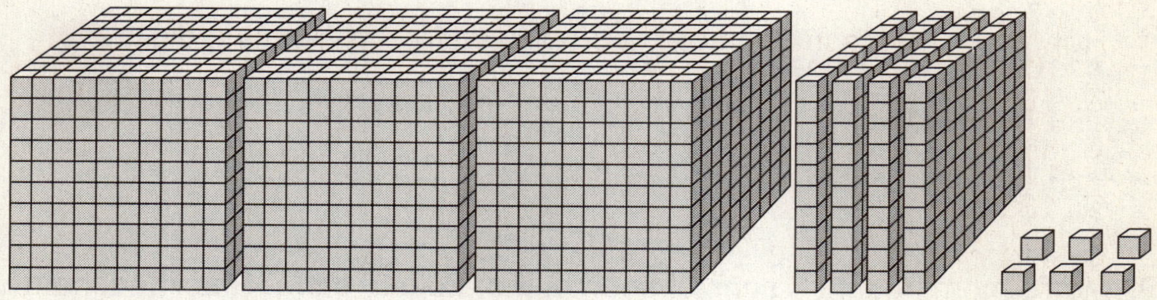

Ⓐ 3,406
Ⓑ 3,460
Ⓒ 4,316
Ⓓ 6,403

5. How is 9,230 written in expanded form?

Ⓐ 9,000 + 200 + 30
Ⓑ 9,000 + 2 + 30
Ⓒ 9,000 + 200 + 3
Ⓓ 9,000 + 2 + 3 + 0

6. Which number is even?

Ⓐ 225
Ⓑ 308
Ⓒ 357
Ⓓ 449

7. How is 608 written in word form?

Ⓐ sixty-eighty
Ⓑ sixty-eight
Ⓒ six hundred eighty
Ⓓ six hundred eight

8. Which of the following is a way of showing 107?

   Ⓐ 17 tens

   Ⓑ 10 tens and 7 ones

   Ⓒ 1 hundred and 7 tens

   Ⓓ 1 ten and 7 ones

9. Matt's mom bought 8 cans of soup at the grocery store. She lined them up on the counter.

   What kind of soup is in the 5th position?

   Ⓐ tomato

   Ⓑ bean

   Ⓒ chicken

   Ⓓ beef

10. Count by 10s. What are the next three numbers?

    50, 60, 70, _____, _____, _____

    Ⓐ 75, 80, 85

    Ⓑ 80, 90, 100

    Ⓒ 90, 100, 110

    Ⓓ 75, 100, 125

Objectives: N.A.3, N.A.4

Lesson 2: Ordering and Comparing Whole Numbers

Numbers have an order. The number 2 always goes after 1 and before 3. You use this order to count and compare numbers. In this lesson, you will order and compare numbers.

Ordering Numbers

The following chart shows some of the counting numbers in order from 1 to 100.

1	2	3	4	5	6	7	8	9	10
11	12	13	14	15	16	17	18	19	20
21	22	23	24	25					
					96	97	98	99	100

Now it's your turn to practice! Fill in the rest of the chart above.

You can use the chart on the previous page to help you order numbers.

Example

What are the next 3 numbers **after** 39?

39, **40**, **41**, **42**

The next 3 numbers after 39 are **40**, **41**, and **42**.

You can also use a number line to help you order numbers.

Example

What are the next 3 numbers **after** 39?

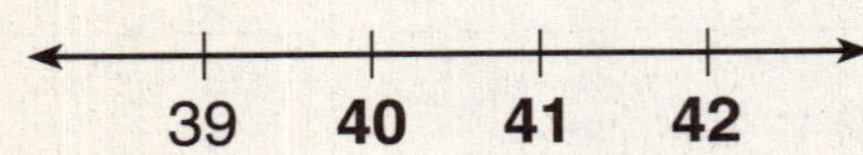

The next 3 numbers after 39 are **40**, **41**, and **42**.

Sometimes you will need to put numbers that are not in a row in order from **least** to **greatest** or **greatest** to **least**.

Example

Put the following numbers in order from **least** to **greatest**.

48, 20, 82

Since 20 comes first on the chart on the previous page, it is the smallest number. The last number on the chart is 82, so it is the largest.

The numbers in order from least to greatest are 20, 48, 82.

Objectives: N.A.3, N.A.4

Practice

Directions: Use the following number line to answer Numbers 1 through 4.

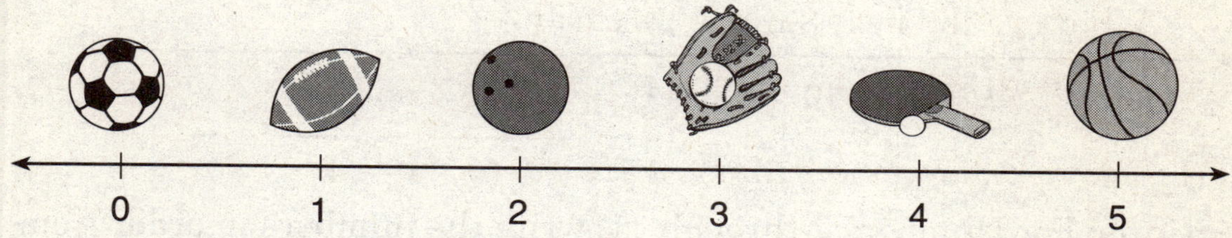

1. Circle the object located at 4.

2. Circle the number where the football is located.

   1 2 3

3. Circle the number where the basketball is located.

   3 4 5

4. Circle the group of objects that is in the same order as those on the number line.

Directions: For Numbers 5 and 6, write the next three numbers.

5. 92, 93, 94, ______, ______, ______

6.

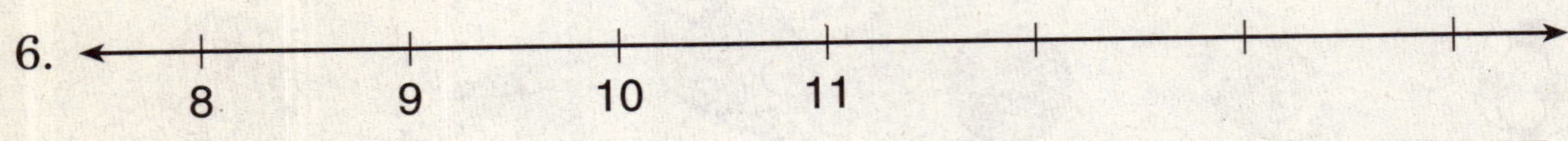

______ ______ ______

Directions: For Numbers 7 through 11, write the numbers in order from **least** to **greatest**.

7. 39, 90, 87 ____________________

8. 28, 24, 27 ____________________

9. 50, 10, 5 ____________________

10. 115, 176, 130 ____________________

11. 208, 280, 199 ____________________

Directions: For Numbers 12 through 16, write the numbers in order from **greatest** to **least**.

12. 54, 50, 61 ____________________

13. 68, 30, 40 ____________________

14. 8, 80, 38 ____________________

15. 140, 132, 190 ____________________

16. 350, 405, 349 ____________________

Objectives: N.A.3

Comparing Numbers

When you compare numbers, you will see words like these:

greater	greatest	smaller	smallest
larger	largest	less	least
bigger	biggest	equal to	the same

You can use these symbols to compare numbers.

$>$ means **is greater than**

$<$ means **is less than**

$=$ means **is equal to**

Think of $>$ **(is greater than)** as the open mouth of a hungry fish. The fish will *always* swim to the bigger number and eat it.

The fish will swim to the bigger number: 10.

Because 10 **is greater than** 9, write **10** $>$ **9**.

You can also think of $<$ **(is less than)** as the open mouth of a hungry fish.

12 17

Again, the fish will swim to the bigger number: 17.

Because 12 **is less than** 17, write **12** $<$ **17**.

You can use a place-value table to help you compare numbers.

Example

Which number is **greater**, 19 or 16?

Tens	Ones
1	9
1	6
↑ same	↑ different

Look at the tens place first. The digits are the same.

Look at the ones place next. The 9 is greater than the 6. The number that has the 9 in the ones place is the greater number. So, 19 is greater than 16, or 19 > 16.

Example

Which number is **greater**, 1,848 or 1,829?

Thousands	Hundreds	Tens	Ones
1	8	4	8
1	8	2	9
↑ same	↑ same	↑ different	

Look at the thousands place first. Both numbers have a 1 in the thousands place.

Look at the hundreds place next. Both numbers have an 8 in the hundreds place.

Look at the tens place. The 4 is greater than the 2. The number that has the 4 in the tens place is the greater number. So, 1,848 is greater than 1,829, or 1,848 > 1,829.

Practice

Directions: For Numbers 1 through 8, compare the numbers. Write the correct symbol on each blank.

>	is greater than
<	is less than
=	is equal to

1. 75 __________ 63

2. 51 __________ 51

3. 199 __________ 187

4. 511 __________ 155

5. 38 __________ 83

6. 634 __________ 634

7. 2,401 __________ 2,410

8. 1,906 __________ 1,899

Directions: Use the following place-value table to answer Numbers 9 and 10.

Thousands	Hundreds	Tens	Ones
3	9	6	7
3	9	7	6

9. Compare 3,967 and 3,976. What is the highest place value in which the numbers are different?

    ____

10. Which number is greater: 3,967 or 3,976? __________

Mathematics Practice

1. Thomas has 430 pennies, Sami has 387 pennies, Wayne has 330 pennies, and Lori has 419 pennies. Who has the **greatest** number of pennies?

   Ⓐ Thomas
   Ⓑ Sami
   Ⓒ Wayne
   Ⓓ Lori

2. What are the next 3 numbers after 43?

   Ⓐ 43, 44, 45
   Ⓑ 42, 41, 40
   Ⓒ 45, 46, 47
   Ⓓ 44, 45, 46

3. Which pair of numbers is compared correctly?

   Ⓐ $63 > 65$
   Ⓑ $16 = 24$
   Ⓒ $92 > 92$
   Ⓓ $54 < 61$

4. What number belongs in the circle on the following number line?

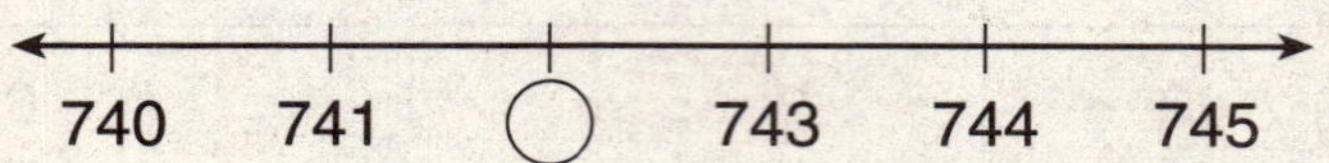

   Ⓐ 740
   Ⓑ 742
   Ⓒ 751
   Ⓓ 752

5. What number comes **before** 56?

   Ⓐ 58
   Ⓑ 56
   Ⓒ 55
   Ⓓ 57

6. The place-value table below shows the populations of three towns.

Thousands	Hundreds	Tens	Ones
2	6	8	0
2	0	8	6
2	8	0	6

   Which of the following correctly compares the numbers?

   Ⓐ $2,680 < 2,086$
   Ⓑ $2,086 > 2,806$
   Ⓒ $2,806 < 2,680$
   Ⓓ $2,680 < 2,806$

7. Which list shows the numbers in order from **least** to **greatest**?

   Ⓐ 88, 78, 98
   Ⓑ 82, 84, 88
   Ⓒ 98, 90, 84
   Ⓓ 88, 80, 90

8. Which number is **less** than 894?

   Ⓐ 849
   Ⓑ 984
   Ⓒ 904
   Ⓓ 897

Lesson 3: Addition

When you **add**, you combine (put together) sets to find how many of something there are altogether. In this lesson, you will review single- and double-digit addition.

Adding One-Digit Numbers

You can use pictures to show how to add. It is also important to memorize basic addition facts.

Example

How many quarters are there altogether?

+

2 1

To get the answer, combine the sets of numbers.

3

There are 3 quarters altogether.

You can write an **addition number sentence** for this problem. The numbers you add are called **addends**. These are the **parts**. The answer is called the **sum**. This is the **whole**.

$2 + 1 = 3$ ← **sum**

↑ ↑

addends

TIP: When you add, the sum is always larger than either of the addends.

Objectives: N.B.1, N.C.1

Practice

Directions: For Numbers 1 and 2, write the number of objects on each blank under the pictures. Add to find how many objects there are altogether.

1. 

________ + ________ = ________

2. 

________ + ________ = ________

Directions: For Numbers 3 through 12, find the sum.

3. 9 + 7 = ________

4. 8 + 5 = ________

5. 7 + 6 = ________

6. 4 + 3 = ________

7. 2 + 9 = ________

8. 9 + 8 = ________

9. 5 + 4 = ________

10. 1 + 8 = ________

11. 6 + 3 = ________

12. 2 + 1 = ________

Adding Two-Digit Numbers

Sometimes you will need to add larger numbers. One way to add two-digit numbers is to use models.

Example

Add: 45 + 51

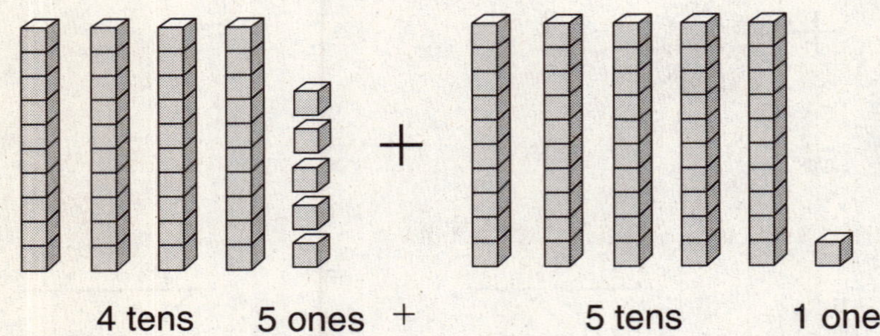

Step 1: **Add the ones.**

5 ones + 1 one = 6 ones

Step 2: **Add the tens.**

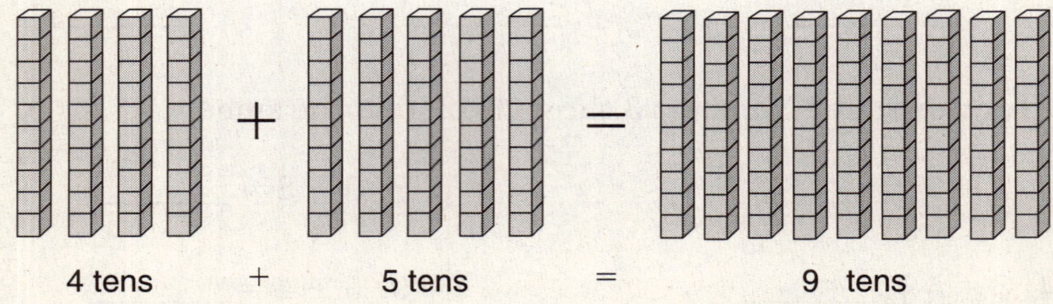

The sum is 96. (There are 96 blocks altogether.)

Objectives: N.C.2

You can also find the sum without using models. First line up the place values. Then add, starting with the ones place.

Example

Add: 45 + 51

Line up the ones places and the tens places. Then add the ones place.

$$\begin{array}{r} 45 \\ +\ 51 \\ \hline 6 \end{array}$$

Now move to the tens place and add.

$$\begin{array}{r} 45 \\ +\ 51 \\ \hline 96 \end{array}$$

The sum is 96. (45 + 51 = 96)

Practice

Directions: For Numbers 1 and 2, draw a model of each number to help you find the sum.

1. 28 + 31

2. 53 + 24

Directions: For Numbers 3 through 10, find the sum.

3. $\begin{array}{r} 12 \\ +\ 47 \\ \hline \end{array}$

4. $\begin{array}{r} 33 \\ +\ 54 \\ \hline \end{array}$

5. $\begin{array}{r} 13 \\ +\ 62 \\ \hline \end{array}$

6. $\begin{array}{r} 71 \\ +\ 25 \\ \hline \end{array}$

7. $\begin{array}{r} 22 \\ +\ 66 \\ \hline \end{array}$

8. $\begin{array}{r} 35 \\ +\ 52 \\ \hline \end{array}$

9. $\begin{array}{r} 40 \\ +\ 44 \\ \hline \end{array}$

10. $\begin{array}{r} 86 \\ +\ 11 \\ \hline \end{array}$

Objectives: N.C.2

Adding with Regrouping

Sometimes when you add two-digit numbers, you need to regroup. When the numbers in the ones place add up to more than 9, regroup 10 ones as 1 ten.

Example

Add: 18 + 23

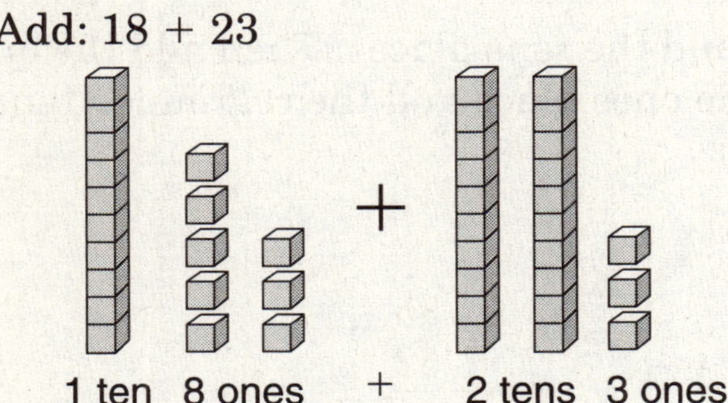

1 ten 8 ones + 2 tens 3 ones

Step 1: **Add the ones.**

8 ones + 3 ones = 11 ones

Step 2: **Regroup the ones when you have more than 9 ones.**

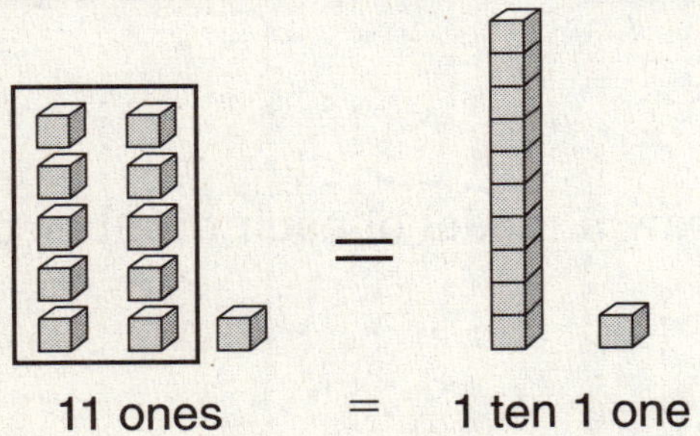

11 ones = 1 ten 1 one

Step 3: **Add the tens.**

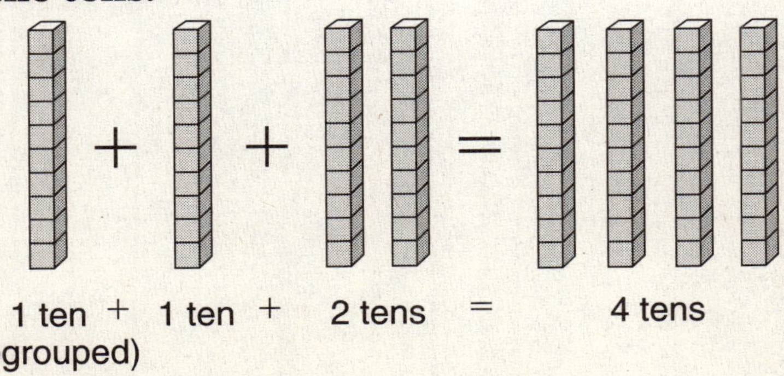

1 ten (regrouped) + 1 ten + 2 tens = 4 tens

The sum is 41. (There are 41 blocks altogether.)

You can also find the sum without using models. This is the way you will add numbers most of the time.

Example

Add: 18 + 23

Line up the ones places and the tens places. Then add the ones place. Write the ones digit in the ones place and the regrouped tens digit above the tens place.

$$\begin{array}{r} \overset{1}{1}8 \\ +\ 23 \\ \hline 1 \end{array}$$

Now move to the tens place and add. Remember to add the 1 ten that was regrouped.

$$\begin{array}{r} \overset{1}{1}8 \\ +\ 23 \\ \hline 41 \end{array}$$

The sum is 41. (18 + 23 = 41)

Practice

Directions: For Numbers 1 and 2, draw a model of each number to help you find the sum.

1. 48 + 16

2. 57 + 37

Objectives: N.C.2

Directions: For Numbers 3 through 10, find the sum. Remember to regroup if there are more than 9 ones.

3. $\begin{array}{r} 68 \\ +\ 24 \\ \hline \end{array}$

4. $\begin{array}{r} 29 \\ +\ 35 \\ \hline \end{array}$

5. $\begin{array}{r} 34 \\ +\ 16 \\ \hline \end{array}$

6. $\begin{array}{r} 54 \\ +\ 17 \\ \hline \end{array}$

7. $\begin{array}{r} 33 \\ +\ 29 \\ \hline \end{array}$

8. $\begin{array}{r} 58 \\ +\ 26 \\ \hline \end{array}$

9. $\begin{array}{r} 42 \\ +\ 38 \\ \hline \end{array}$

10. $\begin{array}{r} 75 \\ +\ 18 \\ \hline \end{array}$

Mathematics Practice

1. What addition number sentence is shown by the following drawing?

Ⓐ 5 + 4 = 9

Ⓑ 4 + 4 = 9

Ⓒ 5 + 5 = 9

Ⓓ 6 + 3 = 9

2. Julia went to Blue Spring State Park to see manatees one weekend while on vacation in Florida. The table below shows how many manatees she saw each day.

Manatees Seen

Saturday	46
Sunday	43

How many manatees did Julia see on the two days combined?

Ⓐ 83

Ⓑ 89

Ⓒ 93

Ⓓ 99

3. Add: 73 + 25

Ⓐ 98

Ⓑ 97

Ⓒ 89

Ⓓ 87

4. Wilda found 8 rocks. Later, she found 4 more. How many rocks did Wilda find altogether?

   Ⓐ 4
   Ⓑ 8
   Ⓒ 10
   Ⓓ 12

5. What is the sum of 39 and 56?

   Ⓐ 115
   Ⓑ 105
   Ⓒ 95
   Ⓓ 85

6. When you add the ones and have to regroup, where do you carry a number?

   Ⓐ to the ones place
   Ⓑ to the tens place
   Ⓒ to the hundreds place
   Ⓓ You don't have to carry.

7. What is 43 + 37?

   Ⓐ 90
   Ⓑ 80
   Ⓒ 70
   Ⓓ 60

Lesson 4: Subtraction

When you **subtract**, you take away something from a set or number. Subtraction is the opposite of addition. In this lesson, you will review single- and double-digit subtraction.

Subtracting One-Digit Numbers

You can use pictures to show how to subtract. It is also important to memorize basic subtraction facts.

Example

Yuna found 5 leaves. Two of them blew away. How many leaves does Yuna have left?

The picture shows that Yuna had 5 leaves. Because 2 blew away, 2 have been taken away (crossed out).

5 take away 2 is 3. Yuna has 3 leaves left.

You can write a **subtraction number sentence** for this problem. The number you subtract from is the **whole**. The number you subtract is one **part**. The **difference** is the answer.

whole → 5 − 2 = 3 ← **difference**
↑
part (points to 2)

TIP: When a number is subtracted from another number, the difference is smaller than the number you started with.

Objectives: N.B.1, N.C.1

Practice

1. Paco had 7 sun stickers. He gave 4 away. How many stickers does Paco have left? Use the following picture to help find the number of stickers Paco has left.

Paco has _____ stickers left.

2. Jamie won 9 tickets playing an arcade game. Keisha won 5 tickets. How many fewer tickets did Keisha win than Jamie?

Keisha won _____ fewer tickets than Jamie.

Directions: For Numbers 3 through 12, find the difference.

3. 13 − 7 = ________

4. 8 − 5 = ________

5. 15 − 6 = ________

6. 5 − 4 = ________

7. 17 − 9 = ________

8. 14 − 9 = ________

9. 10 − 8 = ________

10. 11 − 3 = ________

11. 9 − 2 = ________

12. 7 − 3 = ________

Subtracting Two-Digit Numbers

Sometimes you will need to subtract larger numbers. One way to subtract two-digit numbers is to use models.

Example

Subtract: 24 − 11

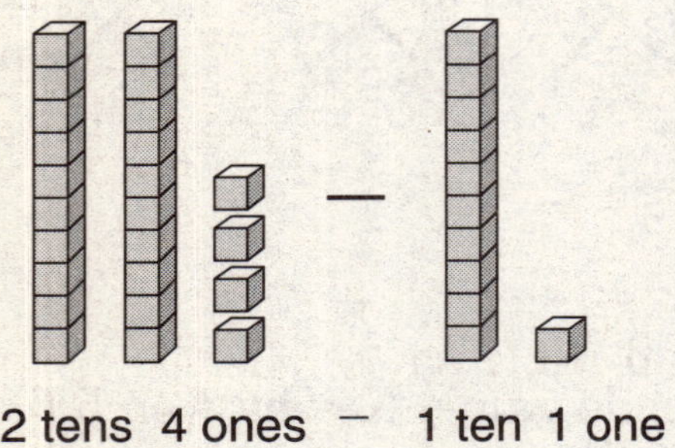

Step 1: **Subtract the ones.**

4 ones − 1 one = 3 ones

Step 2: **Subtract the tens.**

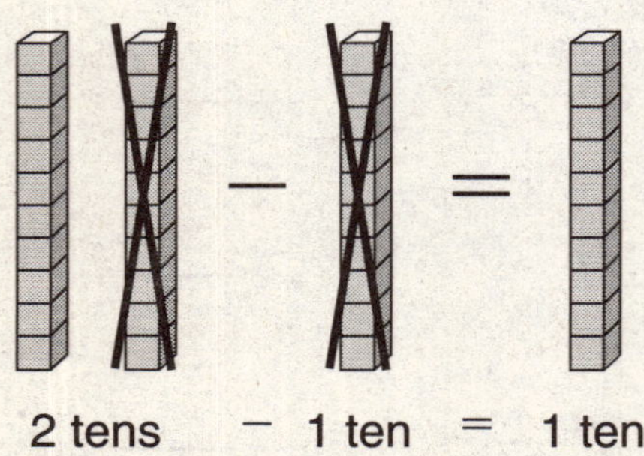

When you take 11 away from 24, you get 13. (There are 13 blocks left.)

Objectives: N.C.2

You can also find the difference without using models. First line up the place values. Then subtract, starting with the ones place.

Example

Subtract: 24 − 11

Line up the ones places and the tens places. Then subtract the ones place.

$$\begin{array}{r} 24 \\ -\ 11 \\ \hline 3 \end{array}$$

Now move to the tens place and subtract.

$$\begin{array}{r} 24 \\ -\ 11 \\ \hline 13 \end{array}$$

The difference is 13. (24 − 11 = 13)

Practice

Directions: For Numbers 1 and 2, draw a model of each number to help you find the difference.

1. 65 − 23

2. 38 − 15

Directions: For Numbers 3 through 10, find the difference.

3. $\begin{array}{r} 89 \\ -\ 12 \\ \hline \end{array}$

4. $\begin{array}{r} 75 \\ -\ 25 \\ \hline \end{array}$

5. $\begin{array}{r} 46 \\ -\ 21 \\ \hline \end{array}$

6. $\begin{array}{r} 97 \\ -\ 35 \\ \hline \end{array}$

7. $\begin{array}{r} 52 \\ -\ 30 \\ \hline \end{array}$

8. $\begin{array}{r} 24 \\ -\ 13 \\ \hline \end{array}$

9. $\begin{array}{r} 71 \\ -\ 41 \\ \hline \end{array}$

10. $\begin{array}{r} 96 \\ -\ 25 \\ \hline \end{array}$

Objectives: N.C.2

Subtracting with Regrouping

Sometimes when you subtract two-digit numbers, you need to regroup. When there are not enough ones to subtract, borrow 1 ten and regroup it as 10 ones.

Example

Subtract: 43 − 18

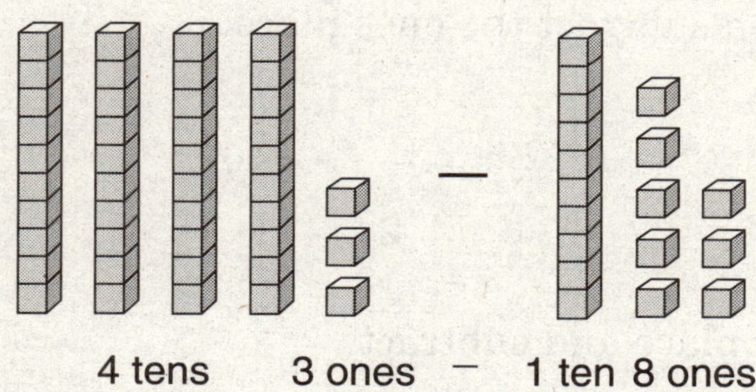

Step 1: **There are not enough ones in the number 43. Borrow a ten and regroup the ones.**

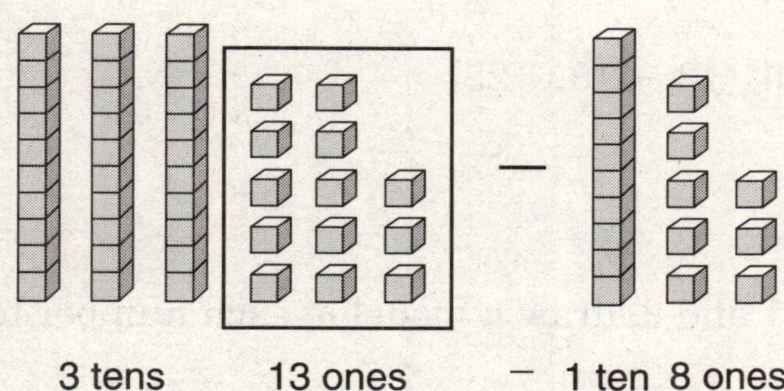

Step 2: **Subtract the ones. Then, subtract the tens.**

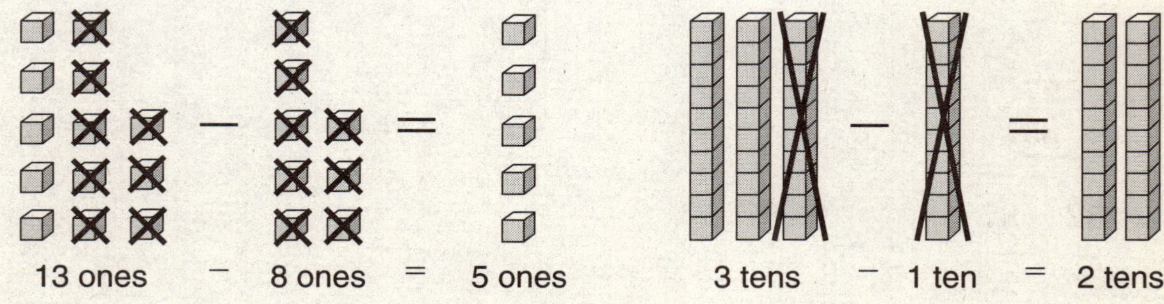

When you take 18 away from 43, you get 25. (There are 25 blocks left.)

You can also find the difference without using models. This is the way you will subtract numbers most of the time.

Example

Subtract: 43 − 18

Line up the ones places and the tens places. Borrow a ten and regroup the ones. Then subtract the ones place.

$$\begin{array}{r} \overset{3}{\cancel{4}}\overset{13}{\cancel{3}} \\ -\ 18 \\ \hline 5 \end{array}$$

Now move to the tens place and subtract.

$$\begin{array}{r} \overset{3}{\cancel{4}}\overset{13}{\cancel{3}} \\ -\ 18 \\ \hline 25 \end{array}$$

The difference is 25. (43 − 18 = 25)

Practice

Directions: For Numbers 1 and 2, draw a model of each number to help you find the difference.

1. 46 − 18

2. 52 − 37

Objectives: N.C.2

Directions: For Numbers 3 through 10, regroup and subtract.

3. $\begin{array}{r} 64 \\ -\ 29 \\ \hline \end{array}$

4. $\begin{array}{r} 92 \\ -\ 35 \\ \hline \end{array}$

5. $\begin{array}{r} 53 \\ -\ 17 \\ \hline \end{array}$

6. $\begin{array}{r} 85 \\ -\ 48 \\ \hline \end{array}$

7. $\begin{array}{r} 70 \\ -\ 51 \\ \hline \end{array}$

8. $\begin{array}{r} 38 \\ -\ 19 \\ \hline \end{array}$

9. $\begin{array}{r} 42 \\ -\ 27 \\ \hline \end{array}$

10. $\begin{array}{r} 94 \\ -\ 45 \\ \hline \end{array}$

Relating Addition to Subtraction

You can use addition to check subtraction and subtraction to check addition. Addition and subtraction are **inverse operations**. That means they are opposites, so one can be used to check the other.

Example

There are 8 rectangles below. If you shade 6 of the rectangles, how many rectangles will **not** be shaded?

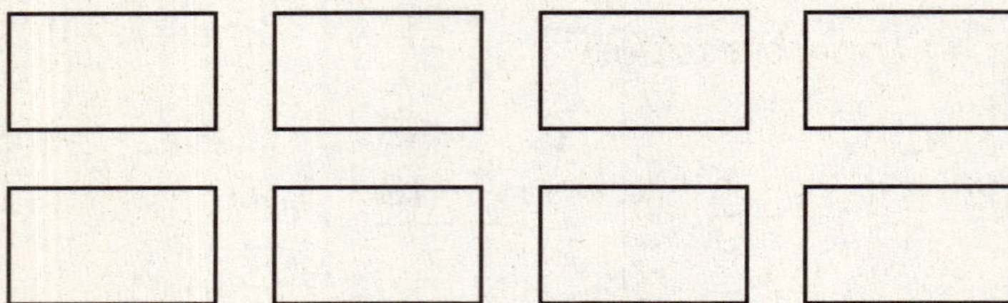

Shade 6 of the rectangles.

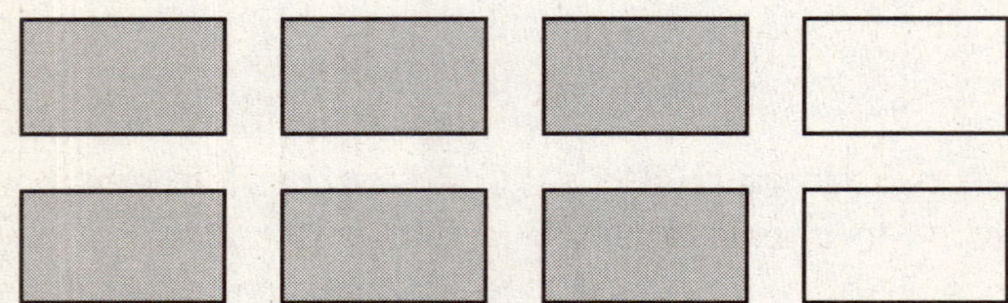

Two of the rectangles are not shaded, so $8 - 6 = 2$.

Is the answer correct? Check by adding **6** to **2**.

$6 + 2 = 8$

Because 8 is the number you started with in the subtraction problem above, the subtraction is correct.

$8 - 6 = 2$ and $6 + 2 = 8$ are **related facts**. They are part of a **fact family**. The other facts in the fact family are $8 - 2 = 6$ and $2 + 6 = 8$.

Objectives: N.B.2

Example

There are 6 boys and 5 girls in Tom's reading group. How many students are in Tom's reading group?

Add: 6 + 5

$$\begin{array}{r} 6 \\ +\ 5 \\ \hline 11 \end{array}$$

Check addition with subtraction.

$$\begin{array}{r} 11 \\ -\ \ 6 \\ \hline 5 \end{array} \quad \text{or} \quad \begin{array}{r} 11 \\ -\ \ 5 \\ \hline 6 \end{array}$$

There are 11 students in Tom's reading group.

Practice

Directions: For Numbers 1 through 8, find the sum or difference. Then write a related number sentence to check that the answer is correct.

1. 18 − 9 = ________

2. 4 + 8 = ________

3. 12 − 5 = ________

4. 3 + 7 = ________

5. 15 − 7 = ________

6. 8 + 6 = ________

7. 13 − 8 = ________

8. 9 + 2 = ________

Mathematics Practice

1. Gilbert has 35 grapes in a bunch. If Gilbert eats 15 of them, how many grapes will be left in the bunch?

   Ⓐ 25

   Ⓑ 20

   Ⓒ 15

   Ⓓ 10

2. Jeremy has earned 9 stars. Lucas has earned 17 stars. How many fewer stars has Jeremy earned than Lucas?

   Ⓐ 6

   Ⓑ 7

   Ⓒ 8

   Ⓓ 9

3. What is 82 − 49?

   Ⓐ 47

   Ⓑ 43

   Ⓒ 37

   Ⓓ 33

4. Mya solved the math problem below.

   8 − 3 = 5

   Which number sentence could Mya use to check her answer?

   Ⓐ 3 + 8 = 11

   Ⓑ 5 + 3 = 8

   Ⓒ 5 − 3 = 2

   Ⓓ 3 − 5 = 8

5. Steven can take 56 pictures on his digital camera. He has already taken 17 pictures. How many more pictures can Steven take?

   Ⓐ 31
   Ⓑ 39
   Ⓒ 41
   Ⓓ 49

6. Which number sentence is shown by the following drawing?

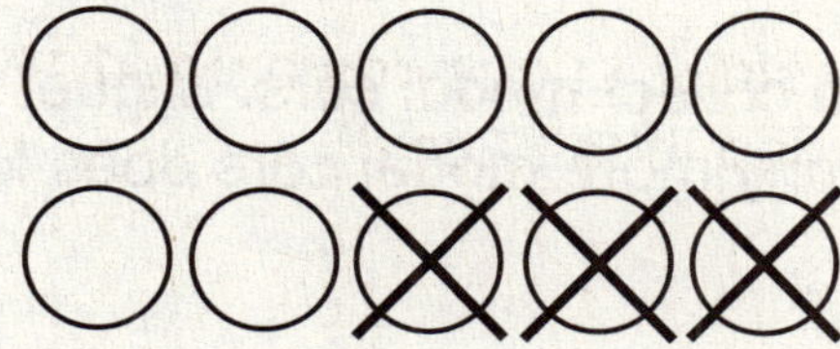

   Ⓐ 10 − 3 = 7
   Ⓑ 10 + 3 = 13
   Ⓒ 13 − 10 = 3
   Ⓓ 7 + 3 = 10

7. Melanie's class has 18 students. There are 7 boys in the class. How many girls are in Melanie's class?

   Ⓐ 8
   Ⓑ 9
   Ⓒ 10
   Ⓓ 11

8. Subtract: 36 − 18

   Ⓐ 18
   Ⓑ 22
   Ⓒ 26
   Ⓓ 28

9. Sylvia solved the math problem below.

   $5 + 9 = 14$

   Which number sentence could Sylvia use to check her answer?

   Ⓐ $14 - 9 = 5$
   Ⓑ $5 - 14 = 9$
   Ⓒ $5 + 14 = 19$
   Ⓓ $9 + 14 = 23$

10. Miguel and Scott both collect model cars. Miguel has 30 model cars, and Scott has 12. How many more model cars does Miguel have than Scott?

    Ⓐ 42
    Ⓑ 28
    Ⓒ 22
    Ⓓ 18

11. What is the difference of 14 and 8?

    Ⓐ 22
    Ⓑ 8
    Ⓒ 6
    Ⓓ 4

12. Subtract: $75 - 23$

    Ⓐ 42
    Ⓑ 51
    Ⓒ 52
    Ⓓ 58

Objectives: N.B.1

Lesson 5: Multiplication and Division

In this lesson, you will review multiplication and division.

Multiplication

When you **multiply**, you add the same amount over and over. Multiplication can be shown by repeated addition, skip counting, or arranging objects in a rectangular array. The numbers that you multiply are called **factors**. The answer when you multiply is called the **product**.

Example

What is the total number of doughnuts on the following plates?

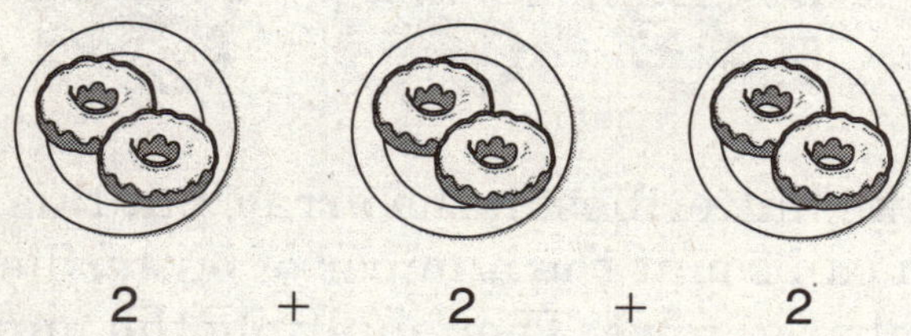

2 + 2 + 2

You can solve this problem using **repeated addition**. Add the number of doughnuts on each plate.

$2 + 2 + 2 = 6$

You can also solve the problem by **skip counting**. Skip count by the number of doughnuts on each plate.

2, 4, 6

There are 2 doughnuts on each plate. There are 3 plates.

You can write this **multiplication number sentence**:

2	×	3	=	6
↑		↑		↑
doughnuts on each plate		**number of plates**		**total number of doughnuts**

There are a total of 6 doughnuts.

You can also use a **rectangular array** to solve a multiplication problem. A rectangular array is a set of objects arranged in the shape of a rectangle.

Example

An art museum has arranged pictures of butterflies in a rectangular array. How many butterflies are shown?

You could count all of the butterflies in the array, but this may take a long time. Instead, you can count the number of butterflies in each row and count the number of rows. Then multiply the numbers.

There are 6 butterflies in each row. There are 3 rows.

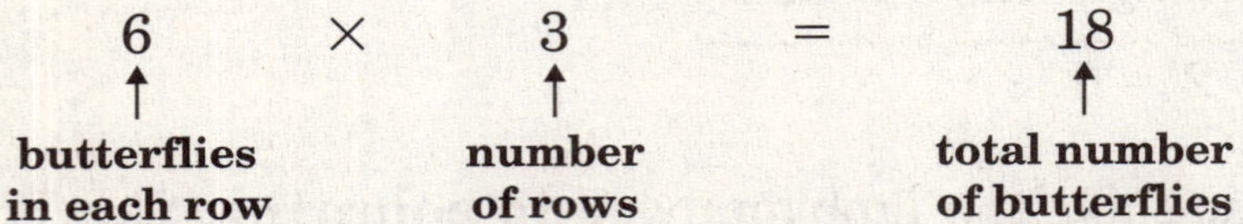

There are 18 butterflies shown.

You can also use repeated addition or skip counting to solve problems arranged in an array. For example: 6 + 6 + 6 = 18 or 6, 12, 18.

TIP: When you multiply whole numbers that are greater than 1, the product is always greater than both of the factors.

Objectives: N.B.1

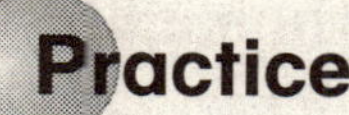

Practice

Directions: Use the following drawing to answer Numbers 1 through 5.

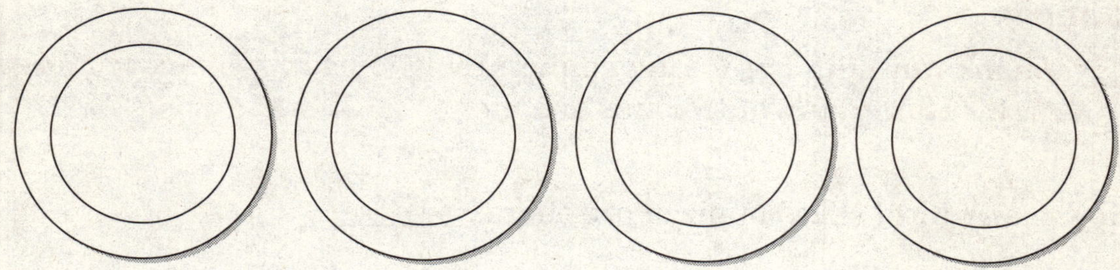

1. Draw 3 cookies on each plate above.

2. How many total cookies are there on the four plates? _______

3. Use repeated addition to show how many cookies there are.

   _______ + _______ + _______ + _______ = _______

4. Use skip counting to show how many cookies there are.

   _______ , _______ , _______ , _______

5. Use multiplication to show how many cookies there are.

   _______ × _______ = _______

6. Look at the following rectangular array of apples. Use repeated addition and multiplication to show how many apples there are.

   _______ + _______ = _______ and _______ × _______ = _______

Directions: Use the following information to answer Numbers 7 through 9.

The Moores ordered 5 pizzas. Each pizza has 4 equal slices.

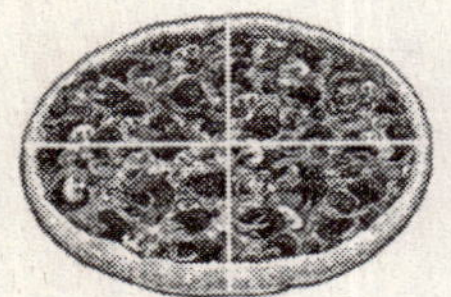 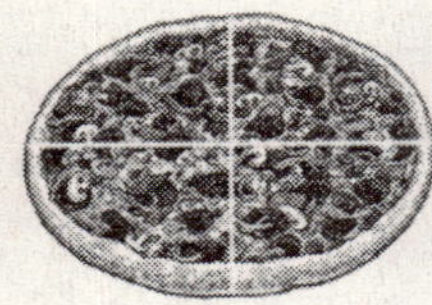 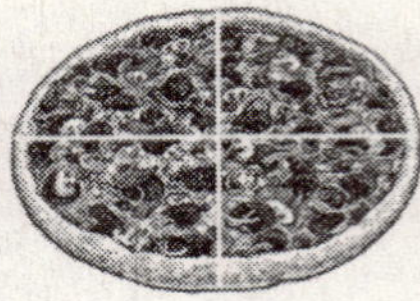 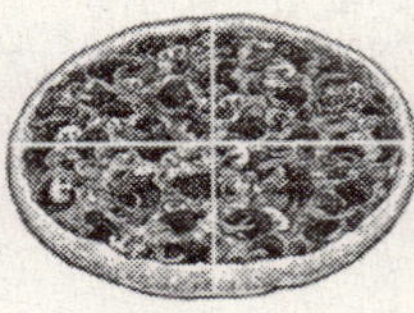 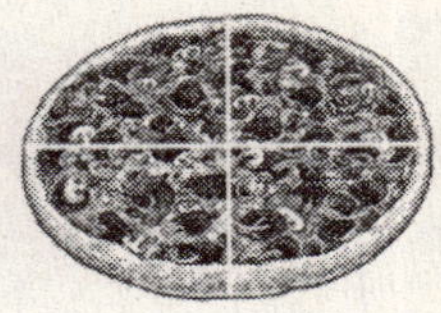

7. How many total slices of pizza are there? _______

8. Use repeated addition to show how many slices there are.

   _______ + _______ + _______ + _______ + _______ = _______

9. Use multiplication to show how many slices there are.

   _______ × _______ = _______

10. The second graders set up these chairs so their parents could watch the class play.

Use skip counting and multiplication to show how many chairs there are.

_______, _______, _______, _______ and _______ × _______ = _______

Objectives: N.B.1

Division

When you **divide**, you make smaller, equal groups of objects from a large group. The number you are dividing is called the **dividend**. The number you are dividing by is called the **divisor**. The answer when you divide is called the **quotient**.

When you know the number in each group, division can be modeled by **repeated subtraction**. Start with the total number of objects and subtract the number of objects in each group until you reach 0.

Seth has 12 roses. He put the roses into groups of 4. How many groups of roses are there?

You can solve this problem using repeated subtraction. Start with the total number of roses, 12. Subtract the number of roses in each group, 4, until you reach 0. Count the number of times you subtracted. This number is the quotient.

$12 - 4 = 8$ (1 time)

$8 - 4 = 4$ (2 times)

$4 - 4 = 0$ (3 times)

You subtracted 4 a total of 3 times.

You can write this **division number sentence**:

12	÷	4	=	3
↑		↑		↑
number of roses		**roses in each group**		**number of groups**

There are 3 groups of roses.

When you know the number of groups, division can be modeled by **sharing equally**.

Example

Sam, Aba, and Becky made 18 sailboats to float on the pond. They shared the sailboats equally among the three of them. How many sailboats did each person get?

Label one sailboat with Sam's name, one with Aba's name, and one with Becky's name. Then start again. Continue until all the sailboats are labeled.

There are 6 sailboats labeled with each person's name.

You can write this **division number sentence**:

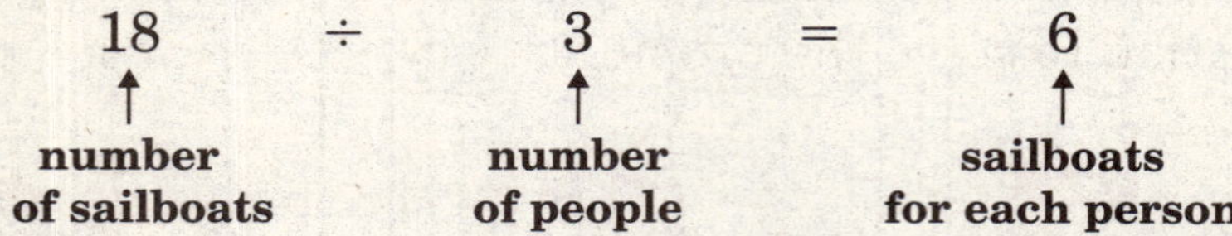

Each person got 6 sailboats.

TIP: When you divide whole numbers, the quotient is always less than the dividend (as long as the divisor is greater than 1).

Objectives: N.B.1

Practice

Directions: Use the following information to answer Numbers 1 and 2.

Mrs. Jackson has 24 students in her class. She divided the students into groups of 6.

1. How many groups of students are there? Use repeated subtraction to solve the problem.

   There are _______ groups of students.

2. Use division to show how many groups of students there are.

   _______ ÷ _______ = _______

Directions: Use the following information to answer Numbers 3 and 4.

Gia is at an amusement park. She only has 45 minutes left before the park closes. She needs 15 minutes to get to each ride and complete it.

3. How many rides does Gia have time to go on before the park closes? Use repeated subtraction to solve the problem.

   Gia has time to go on _______ rides before the park closes.

4. Use division to show how many rides Gia has time to go on before the park closes.

   _______ ÷ _______ = _______

Directions: Use the following information to answer Numbers 5 and 6.

LeBron has 21 sunflower seeds that he wants to plant. He wants to plant them in 7 equal rows.

5. How many sunflower seeds will LeBron plant in each row? Write the row number (1, 2, 3, 4, 5, 6, 7) under each sunflower seed.

LeBron will plant ______ sunflower seeds in each row.

6. Use division to show how many sunflower seeds LeBron will plant in each row.

______ ÷ ______ = ______

Mathematics Practice

1. Cecilia's class saw 12 mountain lions on their field trip to the zoo. There are 3 zookeepers that take care of the mountain lions. Each zookeeper takes care of the same number of mountain lions. How many mountain lions does each zookeeper take care of?

Ⓐ 2
Ⓑ 3
Ⓒ 4
Ⓓ 6

2. Yugi saw 5 starfish at the beach. Each starfish has 5 arms.

Which number sentence can be used to find the total number of arms the starfish have?

Ⓐ $5 + 5 = 10$
Ⓑ $5 \times 5 = 25$
Ⓒ $5 \times 6 = 30$
Ⓓ $25 \div 5 = 5$

3. Which group of objects can be divided into groups of 2?

   Ⓐ

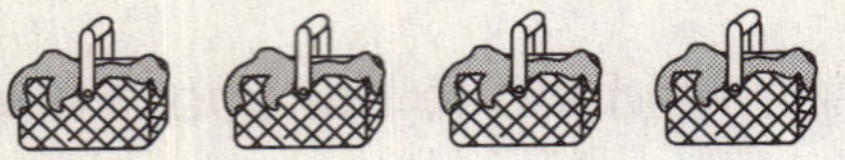

   Ⓑ

   Ⓒ

   Ⓓ

4. Which picture models 4×2?

   Ⓐ

   Ⓑ

   Ⓒ

   Ⓓ

Objectives: N.C.4

Lesson 6: Estimation and Problem Solving

In this lesson, you will learn about different ways of estimating answers. You will also solve story problems.

Estimating

It is a good idea to first **estimate** an answer when you solve problems. When you estimate, you use numbers that are close to the numbers in the problem and are easy to add or subtract. The exact answer to a problem should be close to the estimated answer.

Rounding to the Nearest Ten

Rounding is a quick way to estimate an answer.

The second-grade class read 48 books in one day. Round this number to the nearest ten.

Sometimes it helps to use a number line. 48 is closer to 50 than it is to 40. So, 48 rounded to the nearest ten is 50.

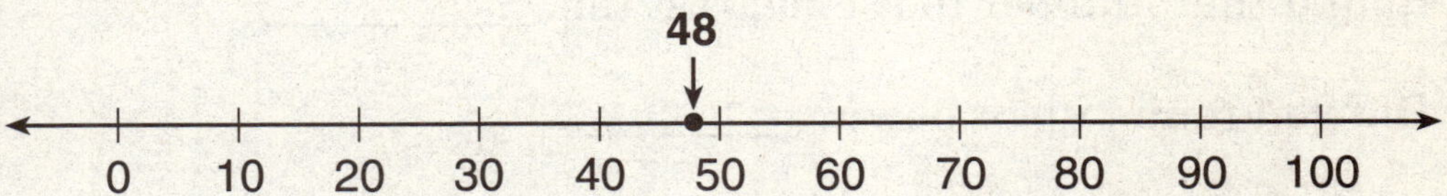

When rounding to the nearest ten, follow these rules:

- If the digit in the ones place is **5 or greater, round up.** ↑

  15 rounded to the nearest ten is 20.

  87 rounded to the nearest ten is 90.

- If the digit in the ones place is **4 or less, round down.** ↓

  62 rounded to the nearest ten is 60.

  14 rounded to the nearest ten is 10.

Example

Estimate: 42 + 87

42 rounded to the nearest ten is 40.

87 rounded to the nearest ten is 90.

40 + 90 = 130

The estimate of 42 + 87 is 130. (The actual sum is 129.)

Practice

1. If you round 32 to the nearest ten, do you round up or down? ___________

2. Round 99 to the nearest ten. ___________

   Did you round 99 up or down? ___________

3. How many students are in your classroom? ___________

   Round that number to the nearest ten. ___________

   Did you round up or down? ___________

4. Estimate: 34 + 29 ___________

5. Estimate: 87 − 41 ___________

6. Estimate: 65 + 38 ___________

7. Estimate: 72 − 23 ___________

Objectives: N.C.4

Front-End Estimation

In **front-end estimation**, you use the front-end digits of the numbers to estimate an answer. The front-end digits are the digits in the greatest place value. Add or subtract only the front-end digits. Treat the other digits as zeros (0).

Example

Use front-end estimation to estimate the following difference.

$$\begin{array}{r} 65 \\ -\ 32 \\ \hline \end{array}$$

Subtract the front-end digits. Treat the other digits as zeros.

$$\begin{array}{r} \downarrow \\ \mathbf{6}5 \\ -\ \mathbf{3}2 \\ \hline ?? \end{array} \qquad \begin{array}{r} \downarrow \\ \mathbf{60} \\ -\ \mathbf{30} \\ \hline \mathbf{30} \end{array}$$

Using front-end estimation, the difference is about 30. (The exact difference is 33.)

Practice

Directions: For Numbers 1 through 4, use front-end estimation to estimate the sum or difference.

1. $\begin{array}{r} 49 \\ +\ 53 \\ \hline \end{array}$

2. $\begin{array}{r} 69 \\ -\ 47 \\ \hline \end{array}$

3. $\begin{array}{r} 45 \\ +\ 38 \\ \hline \end{array}$

4. $\begin{array}{r} 31 \\ -\ 19 \\ \hline \end{array}$

Solving Story Problems

Use the following steps to help you solve story problems.

Step 1: **Understand the problem.**

Don't give up after reading a problem only once! Sometimes a problem will make more sense when you read it a few times. Then you will find out what the problem is asking you to do. It may help to tell yourself the problem in your own words or to draw a picture that shows the problem.

Step 2: **Find what you need.**

Most problems you work with will have numbers. You will need the numbers to solve those problems. Some problems will not have numbers. You will see some problems without numbers later.

Step 3: **Make a plan.**

If you are working with numbers, choose an operation: **addition** (+), **subtraction** (−), **multiplication** (×), or **division** (÷). Write the problem you will solve. If you are working with other types of story problems, you can draw a picture or work backwards to solve the problem.

Step 4: **Estimate an answer.**

Use front-end estimation or rounding to estimate an answer to be sure your actual answer makes sense.

Step 5: **Solve the problem.**

Do the math with the correct operation.

Step 6: **Write a sentence that gives the answer.**

A story problem often needs words in the answer.

Objectives: N.C.4, N.C.5

Example

Hannah paid $26 to go to a softball camp. She paid $12 for a camp T-shirt. How much money did Hannah pay altogether to go to camp and get a T-shirt?

Step 1: **Understand the problem.**

You need to find the amount of money Hannah spent altogether.

Step 2: **Find what you need.**

This story problem has numbers. Here's what you need:

26 (cost of the softball camp)

12 (cost of the camp T-shirt)

Step 3: **Make a plan.**

Most of the time, the word **altogether** means you need to use **addition**. Write the addition problem.

$$\begin{array}{r} 26 \\ +\ 12 \\ \hline \end{array}$$

Step 4: **Estimate an answer.**

Round to the nearest ten to estimate an answer.

$$\begin{array}{r} 26 \\ +\ 12 \\ \hline \end{array} \qquad \begin{array}{r} \mathbf{30} \\ +\ \mathbf{10} \\ \hline 40 \end{array}$$

The answer should be close to 40.

Step 5: **Solve the problem.**

$$\begin{array}{r} 26 \\ +\ 12 \\ \hline 38 \end{array}$$

This is close to the estimate of 40. The answer makes sense.

Step 6: **Write a sentence that gives the answer.**

Hannah paid $38 to go to softball camp and buy a T-shirt.

Practice

Directions: For Numbers 1 through 6, fill in the blanks to solve the following problem.

The Tigers scored a total of 84 points in a basketball game. They scored 45 points before halftime. How many points did the Tigers score after halftime?

1. **Understand the problem.**

   What do you need to find?

   ____

2. **Find what you need.**

   Total points scored ________ Points scored before halftime ________

3. **Make a plan.**

4. **Estimate an answer.**

   The answer should be close to ________.

5. **Solve the problem.**

   Is the answer close to the estimate in Number 4? ________

6. **Write a sentence that gives the answer.**

   ____

Objectives: N.C.4, N.C.5

Directions: For Numbers 7 through 10, set up a number sentence and solve it. Then write a sentence that gives the answer.

7. Ramon has 24 seashells. His sister Rene has 47 seashells. How many seashells do Ramon and Rene have altogether?

8. Samantha counted 92 birds while walking through the park. Brady counted 67 birds. How many more birds did Samantha count than Brady?

9. There are 50 states in the United States of America. Of those states, 17 begin with a letter from A through K. How many states begin with a letter from L through Z?

10. At lunch today, there were 37 second graders and 48 third graders in the cafeteria. How many second and third graders were in the cafeteria altogether?

Problem-Solving Strategies

You cannot solve every problem using number sentences. Two problem-solving strategies are to draw a picture or work backwards.

Draw a Picture

Pictures or drawings can help you put information together.

Nate, Meg, and Peg are waiting in line in front of the classroom door. Peg is the last one in line. Meg is in front of Nate. Who is closest to the door? Who is farthest away from the door?

Draw a picture to help you. You know that Peg is last in line.

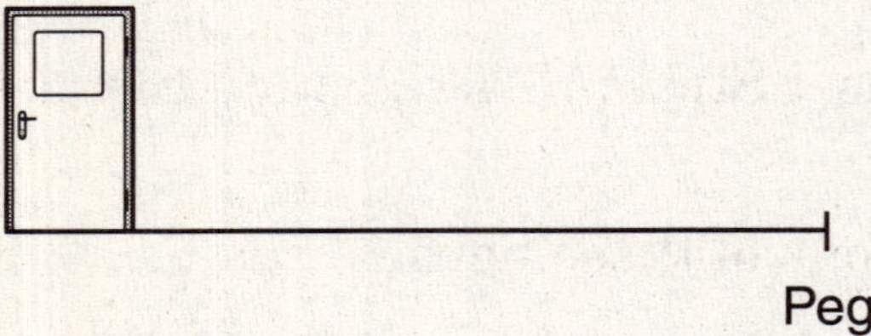

Now show that Meg is in front of Nate.

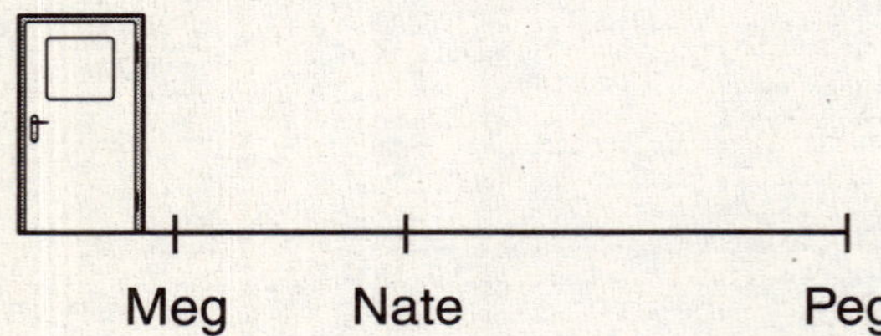

The drawing helps you see who is closest to the door.

Meg is closest to the door. Peg is farthest away from the door.

Objectives: N.C.3

Practice

Directions: For Numbers 1 and 2, draw a picture to help you solve the problem.

1. Three students are lined up from shortest to tallest.

   Dustin is the shortest.

   David is taller than Sergio.

   Who is the tallest? ____________________

2. There are three houses on Sunset Street. Trace around the dashed lines to help you see each house.

   Ted, Sue, and J.T. each live in one of these houses.

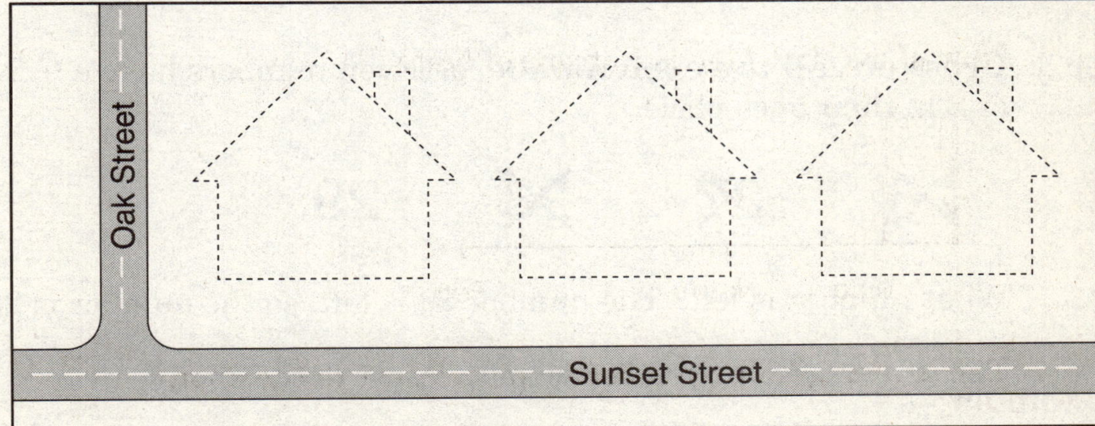

   Sue does not live next to Ted.

   Ted lives on the corner of Sunset Street and Oak Street.

   Who lives in the middle house? ____________________

Work Backwards

Sometimes you will know the total or final answer to a problem, but one of the pieces will be missing. When this happens, you can work backwards to find the missing piece.

Example

Miss Gale's class likes number puzzles. They need to use the following hints to find a number.

The number is odd. The number is between 25 and 30. The number is greater than 27. What is the number?

Step 1: **List all the numbers between 25 and 30.**

26 27 28 29

Step 2: **Cross out 27 and all the numbers that are less than 27.**

~~26~~ ~~27~~ 28 29

Step 3: **Cross out all the even numbers.** (Even numbers have a 0, 2, 4, 6, or 8 in their ones place.)

~~26~~ ~~27~~ ~~28~~ 29

What number is left? The number 29 is left, so the number is 29.

Example

Noah had 2 gumballs. He bought some more at the store. Now he has 5 gumballs. How many gumballs did Noah buy?

Start with the 5 gumballs Noah has now. Take away the 2 gumballs he had to begin with.

Noah bought 3 gumballs at the store.

Practice

1. Joey and Francis found this clue on a school scavenger hunt.

The final prize is in a locker.
The number is odd.
The number is greater than 26.
The number is between 20 and 29.

Which locker is the final prize in?

Step 1: **Fill in the blanks with the numbers between 20 and 29.**

20 _____ _____ _____ _____

_____ _____ _____ _____ 29

Step 2: **Cross out 26 and all of the numbers less than 26.**

Step 3: **Cross out all of the even numbers.**

What number is left? ________________

The prize is in locker number ________________.

2. Juan had 6 toy cars. After opening presents at his birthday party, he has 15 toy cars. How many toy cars did Juan get for his birthday?

Juan got __________ toy cars for his birthday.

Guess and Check

Sometimes you may want to guess at an answer. After you guess, check to see if your answer is correct.

Example

About how many trees are in the park?

Take a guess from one of the following. Circle your guess.

5 15 25

Now find the exact number by counting the trees.

There are _________ trees.

Example

What is the missing number?

3 + _____ = 12

Try 7. 3 + 7 = 12? No, 3 + 7 = 10.

Try 8. 3 + 8 = 12? No, 3 + 8 = 11.

Try 9. 3 + 9 = 12? Yes!

The missing number is 9.

Objectives: N.C.3

Practice

1. Guess the number of stars that are shown below. ______________

Now find the exact number by counting the stars. ______________

2. Use guess and check to find the missing number.

   13 + _____ = 29

   The missing number is _________.

3. Felipe bought two of the items that are shown below. He spent 77¢ on the two items.

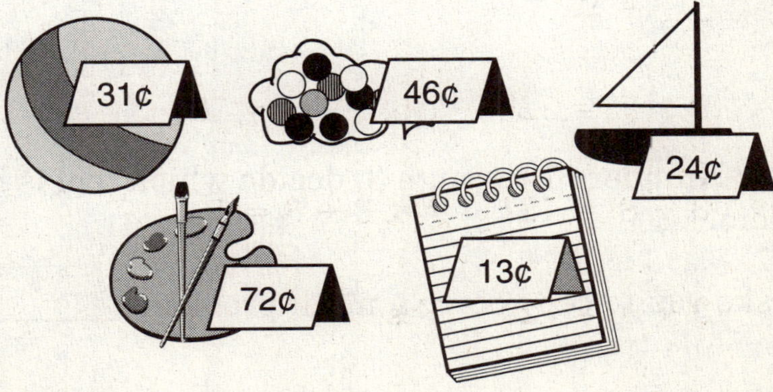

Use guess and check to find the two items that Felipe bought.

____

Which Tool Is Best to Use?

Once you choose a way to solve a problem, you also choose a tool to use. The tool you choose to solve a problem depends on what type of problem you need to solve.

When you have a lot of numbers, use a calculator.

Watch closely as you press numbers. Sometimes people think that a calculator can't be wrong, but it only does what you tell it to do. It's up to you to tell the calculator the correct things to do.

When you need a quick answer, use your memory.

Addition and subtraction facts are good to remember when you need a quick answer. Sometimes it seems really hard to remember all that stuff—it takes practice.

When you need to write down ideas or draw a picture, use a pencil and paper.

You can use a pencil and paper to solve most problems.

Practice

Directions: For Numbers 1 through 3, decide which tool is best to use to solve the problem.

1. An adult asks you to copy down a math problem. ____________________

2. You add the scores from 23 different video games. ____________________

3. You take a timed test on addition facts. ____________________

Mathematics Practice

1. Maria has 38 total marbles. She has 15 black marbles. The rest of the marbles are red. How many red marbles does Maria have?

   Ⓐ 23
   Ⓑ 33
   Ⓒ 43
   Ⓓ 53

2. What number do you get if you round 54 to the nearest ten?

   Ⓐ 60
   Ⓑ 50
   Ⓒ 40
   Ⓓ 5

3. There are three shapes in a row. The square is after the circle. The triangle is before the circle. What is the order of the shapes?

   Ⓐ circle, triangle, square
   Ⓑ square, triangle, circle
   Ⓒ square, circle, triangle
   Ⓓ triangle, circle, square

4. Estimate: 66 − 38

   Ⓐ 10
   Ⓑ 20
   Ⓒ 30
   Ⓓ 40

5. Christina's favorite number is odd. The digits of the number add to 4. The number is between 20 and 40. What is Christina's favorite number?

Ⓐ 22

Ⓑ 31

Ⓒ 37

Ⓓ 40

6. Jordan swam for 48 minutes. Kyoko swam for 29 minutes. How many minutes longer did Jordan swim than Kyoko?

Ⓐ 19 minutes

Ⓑ 21 minutes

Ⓒ 48 minutes

Ⓓ 77 minutes

7. Use front-end estimation to estimate the following sum.

$$\begin{array}{r} 23 \\ +\ 56 \\ \hline \end{array}$$

Ⓐ 40

Ⓑ 50

Ⓒ 70

Ⓓ 80

8. Which group of numbers has a sum of 80?

Ⓐ 13 + 6 + 48 + 23

Ⓑ 35 + 17 + 17 + 20

Ⓒ 24 + 11 + 3 + 36

Ⓓ 9 + 4 + 15 + 52

Objectives: N.A.7

Lesson 7: Fractions

Fractions are numbers between whole numbers. They name parts of a whole. The **numerator (top number)** of the fraction tells how many parts of the whole are special in some way. The **denominator (bottom number)** tells the number of parts the whole is divided into.

$$\begin{array}{r} \textbf{numerator} \rightarrow \\ \textbf{denominator} \rightarrow \end{array} \frac{2}{5}$$

Words

The word form of a fraction is the same as how you say the fraction out loud. It is the number in the numerator followed by the word for the denominator. The following table shows the words for denominators of 2, 3, 4, 6, 8, and 10.

Denominator	Word Name
2	half (halves)
3	third(s)
4	fourth(s)
6	sixth(s)
8	eighth(s)
10	tenth(s)

Example

What are the word forms of $\frac{1}{2}$, $\frac{1}{3}$, $\frac{3}{4}$, $\frac{2}{6}$, $\frac{5}{8}$, and $\frac{9}{10}$?

$\frac{1}{2}$: one half

$\frac{1}{3}$: one third

$\frac{3}{4}$: three fourths

$\frac{2}{6}$: two sixths

$\frac{5}{8}$: five eighths

$\frac{9}{10}$: nine tenths

Parts of a Whole

A whole object can be divided into a number of equal parts. To be a fraction, the parts all have to be the same size. A fraction names the parts of the whole object that are special in some way.

Example

What fraction of the rectangle is shaded?

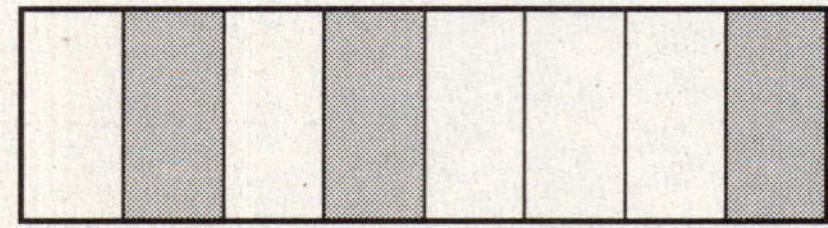

numerator → 3 (parts that are shaded)
denominator → 8 (parts the whole is divided into)

Three eighths, or $\frac{3}{8}$, of the rectangle is shaded.

Fractions can have the same numerator and denominator. When this happens, it means that all of the parts of the whole are special in some way. So, a fraction like this is the same as 1 whole.

Example

What fraction of the rectangle is shaded?

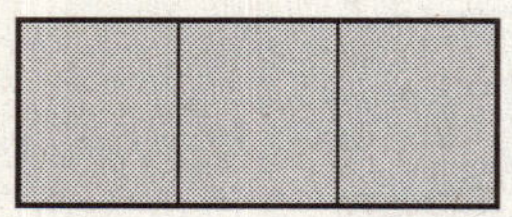

numerator → 3 (parts that are shaded)
denominator → 3 (parts the whole is divided into)

Three thirds, or $\frac{3}{3}$, of the rectangle is shaded. Or, **1 whole** rectangle is shaded.

TIP: The **D**enominator is **D**ownstairs.

Objectives: N.A.7

Practice

1. What is the **word form** of the fraction of the following circle that is shaded?

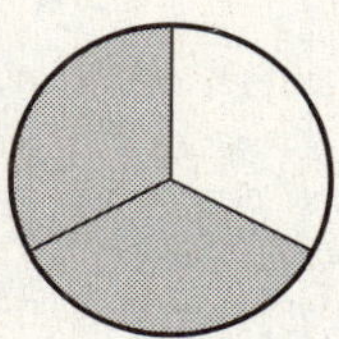

____

2. Shade three fourths of the following circle.

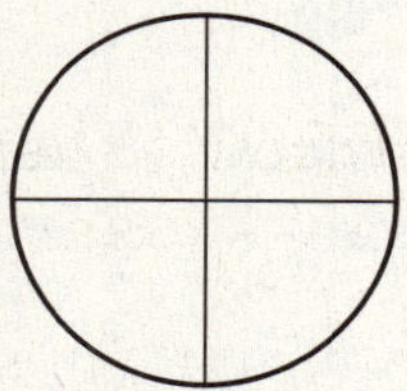

3. Shade $\frac{2}{6}$ of the following rectangle.

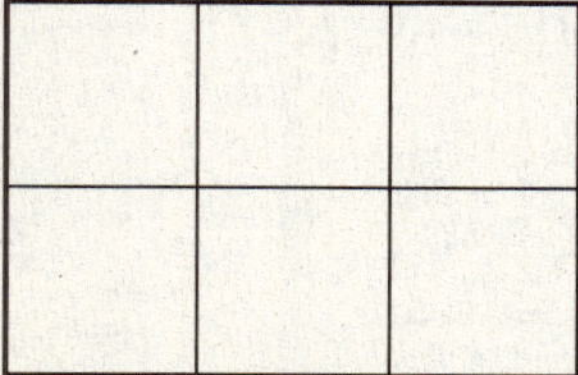

4. What fraction of the following rectangle is shaded?

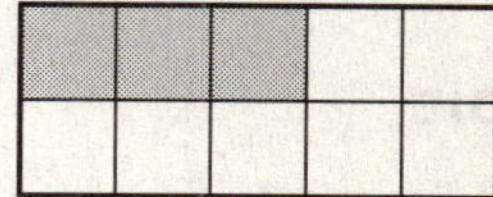

____

5. Which of the following shows $\frac{2}{8}$ of the circle shaded?

A. 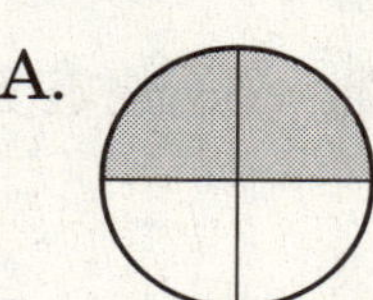

B. 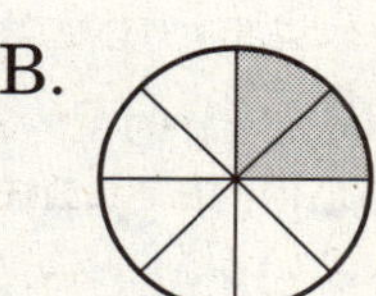

C. 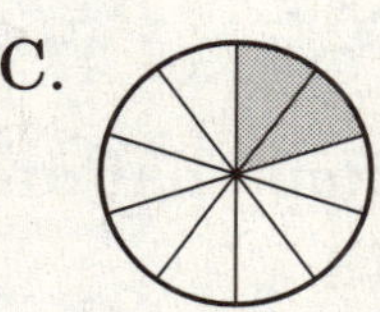

D. 

Parts of a Set

In this lesson, you learned that a fraction can name parts of a whole, like one half of a slice of pizza. A fraction can also name parts of a set, like one out of a set of two lions.

Example

Set of 2
Here is a set of 2 lions.

There is 1 lion that is shaded. **1** out of **2** lions is shaded.

One half, or $\frac{1}{2}$, of the lions are shaded.

Example

Set of 3
Here is a set of 3 lions.

There are 2 lions that are shaded. **2** out of **3** lions are shaded.

Two thirds, or $\frac{2}{3}$, of the lions are shaded.

Example

Set of 4
Here is a set of 4 lions.

There are 4 lions that are shaded. **4** out of **4** lions are shaded.

Four fourths, or $\frac{4}{4}$, of the lions are shaded. Or, the **whole** set of lions is shaded.

Objectives: N.A.7

Practice

Directions: Use the following picture to answer Numbers 1 and 2.

1. What fraction of the fruit are apples? ______

2. What fraction of the fruit are bananas? ______

Directions: For Numbers 3 through 7, write the fraction that names the shaded parts of the set. Then write the word form of the fraction.

3. ______ ____________________

4. ______ ____________________

5. ______ ____________________

6. ______ ____________________

7. ______ ____________________

Comparing Fractions

When you compare fractions, the sizes of the whole must be the same.

Example

Compare the fractions that show the shaded parts of the following circles.

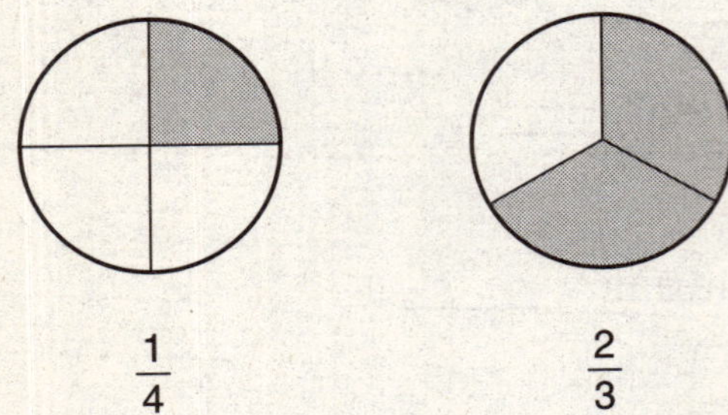

$\frac{1}{4}$ $\frac{2}{3}$

The whole circles are the same size. The shaded parts of the first circle are less than the shaded parts of the second circle.

$\frac{1}{4} < \frac{2}{3}$

Example

Compare the fractions that show the striped parts of the following sets of marbles.

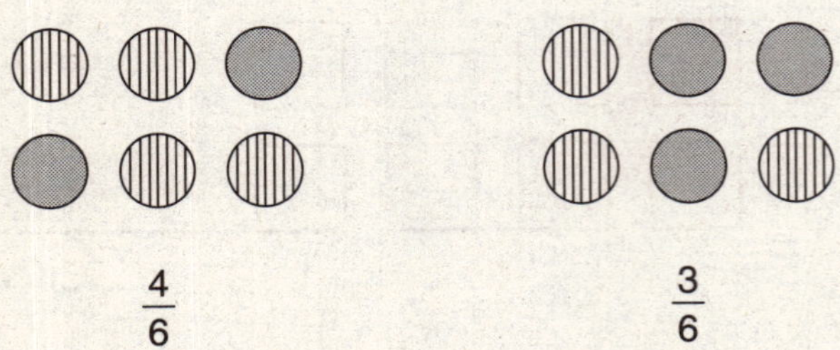

$\frac{4}{6}$ $\frac{3}{6}$

Both sets have a total of 6 marbles. There are 4 striped marbles in the first set and only 3 striped marbles in the second set.

$\frac{4}{6} > \frac{3}{6}$

Objectives: N.A.8

Practice

Directions: For Numbers 1 through 4, write the fraction that shows the shaded parts underneath each object. Then compare the two fractions by writing <, >, or = on the blank between the objects.

1. 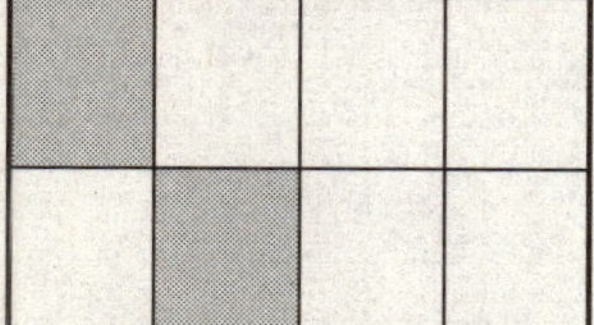

______ ______ ______

2. 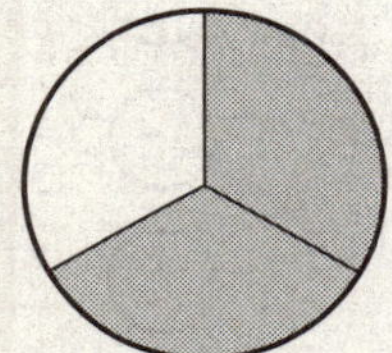 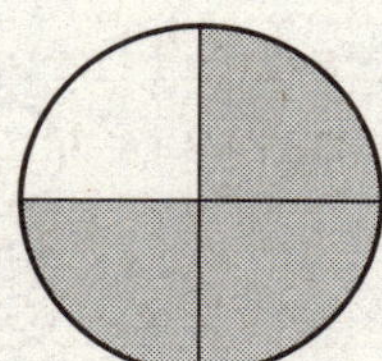

______ ______ ______

3.  

______ ______ ______

4. 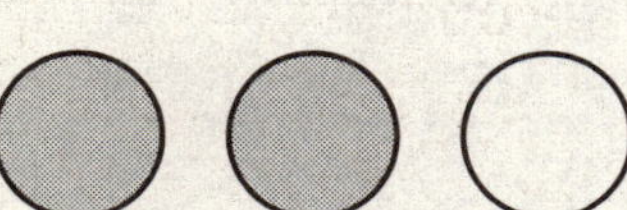 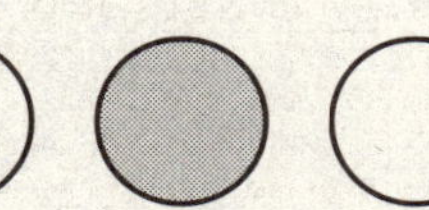

______ ______ ______

Mathematics Practice

1. In word form, the fraction $\frac{10}{10}$ is ten tenths. What is another way to say $\frac{10}{10}$ in word form?

   Ⓐ one tenth

   Ⓑ one half

   Ⓒ one whole

   Ⓓ ten wholes

2. Which set of beads shows $\frac{3}{4}$ that have stripes?

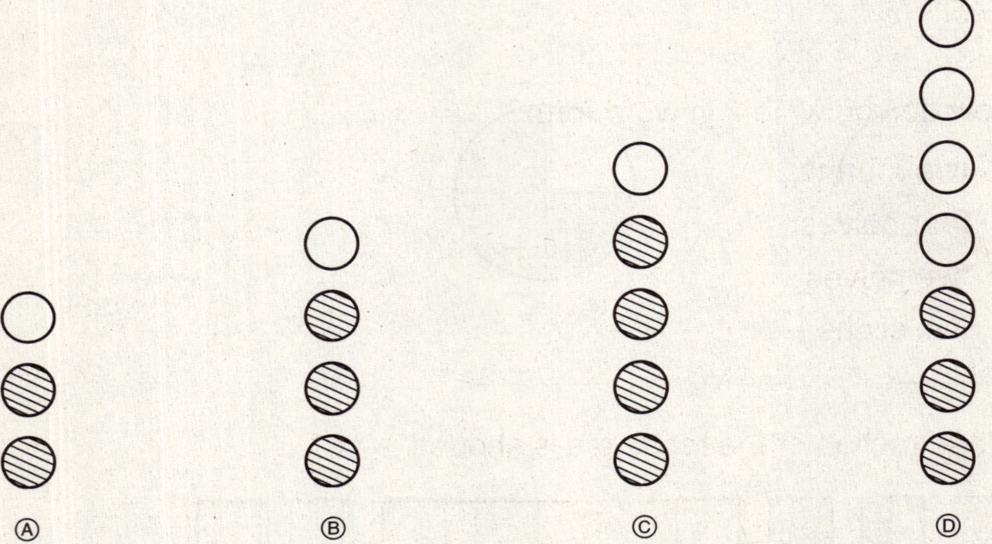

3. Which circle has the **greatest** fraction of shaded parts?

Ⓐ

Ⓑ

Ⓒ

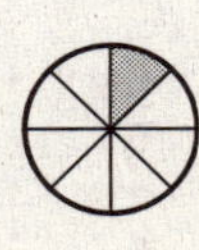

Ⓓ

4. Which correctly compares the shaded parts of the following rectangles?

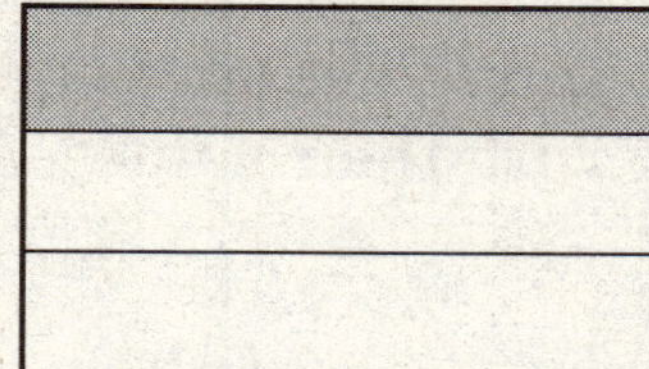

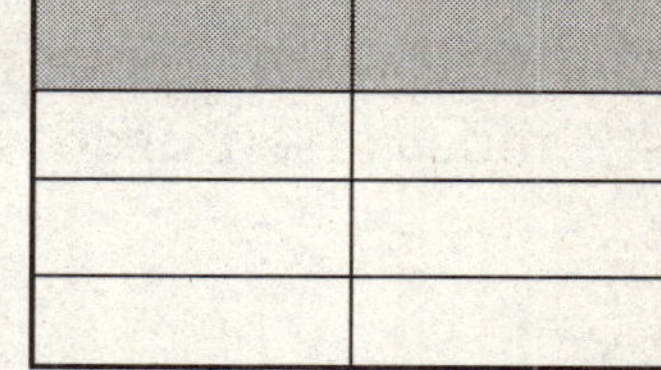

Ⓐ $\frac{1}{3} < \frac{2}{3}$

Ⓑ $\frac{1}{3} > \frac{2}{6}$

Ⓒ $\frac{1}{3} < \frac{2}{8}$

Ⓓ $\frac{1}{3} > \frac{2}{8}$

5. How do you write $\frac{2}{4}$ in word form?

Ⓐ two fourths

Ⓑ four halves

Ⓒ two halves

Ⓓ two sixths

6. What fraction of the rectangle is shaded?

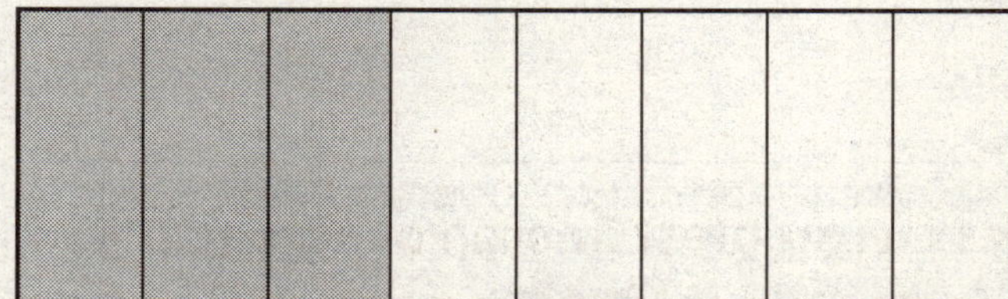

Ⓐ $\frac{5}{3}$

Ⓑ $\frac{5}{8}$

Ⓒ $\frac{3}{5}$

Ⓓ $\frac{3}{8}$

Lesson 8: Money

In this lesson, you will count money amounts. The following drawing shows the coins that are most often used, along with their values. A one-dollar bill is also shown.

penny = 1¢

nickel = 5¢ = 5 pennies

dime = 10¢ = 2 nickels

quarter = 25¢ = 2 dimes and 1 nickel

half dollar = 50¢ = 2 quarters

= 100¢

one dollar
$1.00

Objectives: N.A.9

Writing Money

The value of dollars and cents is shown with symbols or words.

$4 or $4.00 (four dollars)

11¢ or $0.11 (eleven cents)

$4.11 (four dollars and eleven cents)

Counting Money

Counting groups of the same coins is like skip counting.

Use skip counting to find the total value of the nickels.

Each nickel is worth 5¢. This is like skip counting by 5s.

Skip counting shows that 6 nickels equals thirty cents, 30¢, or $0.30.

To count a collection of mixed coins, start with the coin of greatest value. Then continue with coins of lesser value.

Example

What is the total value of the following coins?

Start with the quarter, then the dimes, and finally the pennies. Add the amounts as you go from left to right.

The total value is fifty-nine cents, 59¢, or $0.59.

Equal amounts of money can be shown in different ways.

Example

Each collection of coins has a value of 42¢.

Objectives: N.A.9

To count a collection of bills and coins, first count the bills. Then count the coins.

Example

What is the total value of the following amount of money?

First count the bills. Then count the coins starting with the half dollar, followed by the quarter, dime, nickel, and finally the pennies.

$1.00

$2.00

$2.50 $2.75 $2.85 $2.90 $2.91 $2.92

The total value is two dollars and ninety-two cents, or $2.92.

Practice

Directions: For Numbers 1 and 2, use skip counting to find the value of the coins.

1. 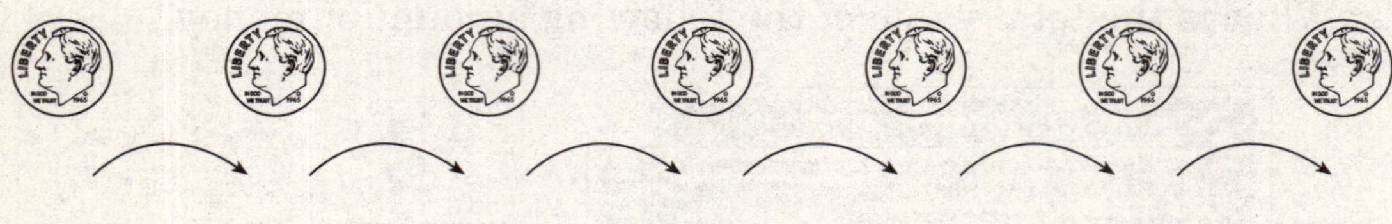

   ______ ______ ______ ______ ______ ______ ______

2. 

   ______ ______ ______ ______ ______ ______ ______

Directions: Use the sets of coins shown for each person to answer Numbers 3 through 5.

Paula

Annette

Steve

Indira

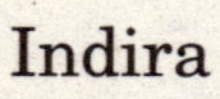

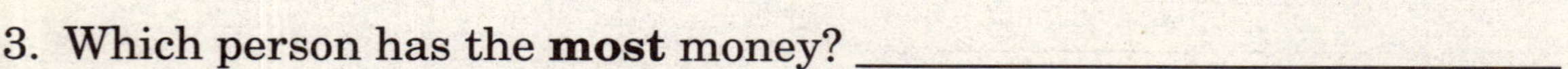

3. Which person has the **most** money? ______________________

4. Which person has the **least** money? ______________________

5. Which person has $0.46? ______________________

Objectives: N.A.9

Directions: For Numbers 6 through 10, count the total value of the amount of money shown. Write the amount using symbols.

6. 

____

7. 

____

8. 

____

9. 

____

10. 

____

11. Circle the coins to show 83¢. (There is more than one way to show this amount.)

Directions: For Numbers 12 and 13, draw coins to show the same amount of money in a different way.

12.

13.

Mathematics Practice

1. Look at the coins below.

50¢ 60¢ 70¢ ______ 85¢ 90¢

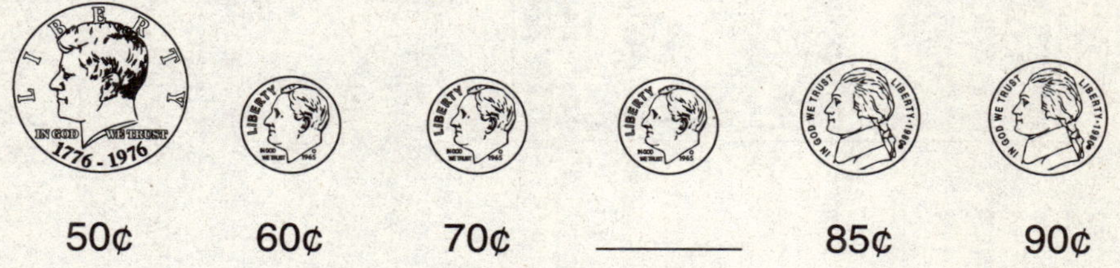

What is the missing value when counting the coins?

Ⓐ 71¢

Ⓑ 75¢

Ⓒ 80¢

Ⓓ 81¢

2. Which set of coins shows the same amount as the coins in the box?

Ⓐ

Ⓑ

Ⓒ

Ⓓ

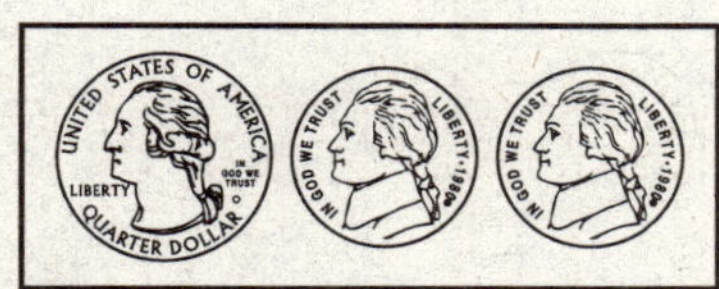

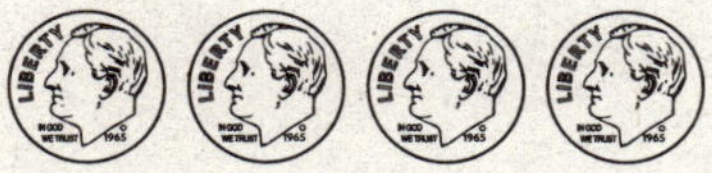

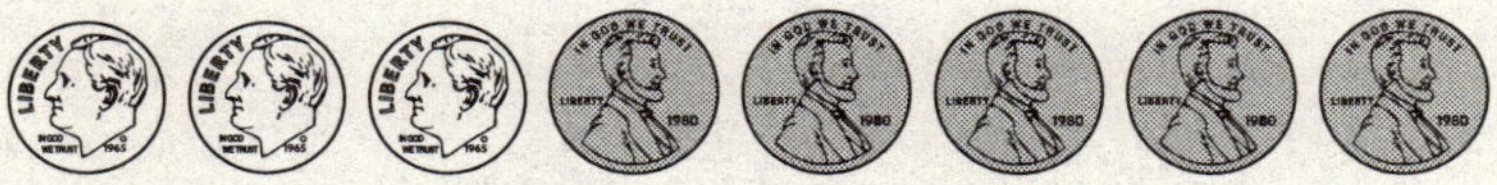

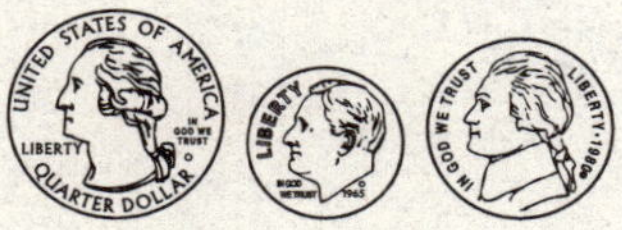

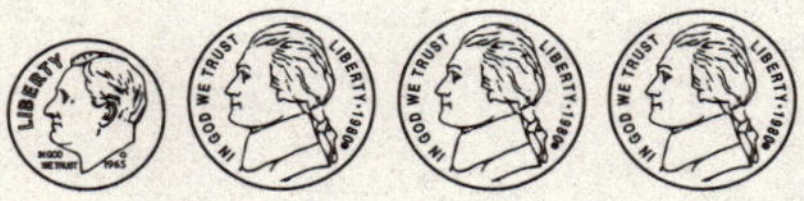

3. How much money is shown below?

Ⓐ $1.95

Ⓑ $2.05

Ⓒ $2.45

Ⓓ $2.55

4. John wants to buy a book that costs 89¢. Which set of coins is enough to buy the book?

Ⓐ

Ⓑ

Ⓒ

Ⓓ

Unit 2

Algebra

On the American flag, the colors of the stripes from top to bottom are red, white, red, white, red, white, red, white, red, white, red, white, and red. The number of stars in each row are 6, 5, 6, 5, 6, 5, 6, 5, and 6.

In this unit, you will work with both shape patterns and number patterns. You will describe, continue, and write rules for the patterns. You will also find missing numbers in open number sentences and describe change.

In This Unit

Lesson 9: Patterns

In this lesson, you will learn about patterns. A **pattern** repeats things or groups of things. There are many different types of patterns.

Shape Patterns

A **shape pattern** repeats circles, squares, triangles, or other shapes.

Example

What are the next two figures in this pattern?

Step 1: **Study the shapes.**

There are circles and triangles in the pattern.

Step 2: **Study the order.**

The order is circle, triangle, circle, triangle, circle.

Step 3: **Study the sizes.**

The sizes are big, big, small, small, big.

Step 4: **Study the shadings or markings.**

There are no shadings or markings.

Here are the next two figures in the pattern.

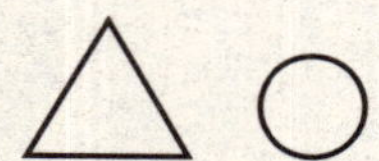

Objectives: A.A.1, A.A.3

Example

Now look at this pattern. What are the next two figures?

This pattern is made up of triangles that are all the same size. The only difference is that every other one is shaded.

Here are the next two figures in the pattern.

To make a shape pattern that is similar to another pattern, follow the same rule, except use different shapes.

Example

Make a pattern that is similar to the triangle pattern in the example above.

The triangle pattern uses triangles that are the same size. Every other triangle is shaded. To make a similar pattern, choose another shape. Make all the shapes the same size, and shade every other one.

The following circle pattern in similar to the triangle pattern.

Practice

Directions: For Numbers 1 and 2, draw the next two figures in each pattern.

1. 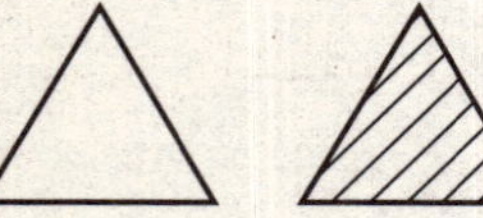 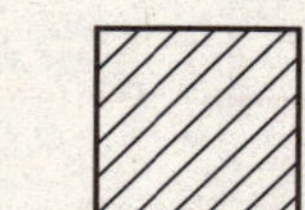 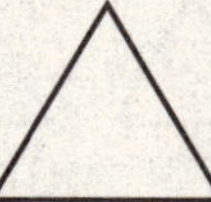

2. 

3. Tony's grandma is making a quilt. Shade the next row of the quilt to follow the pattern.

4. Billy's dad is building a wall of bricks. He noticed the bricks are making a pattern as the wall gets taller. Draw in the next two rows of bricks in the space above the wall.

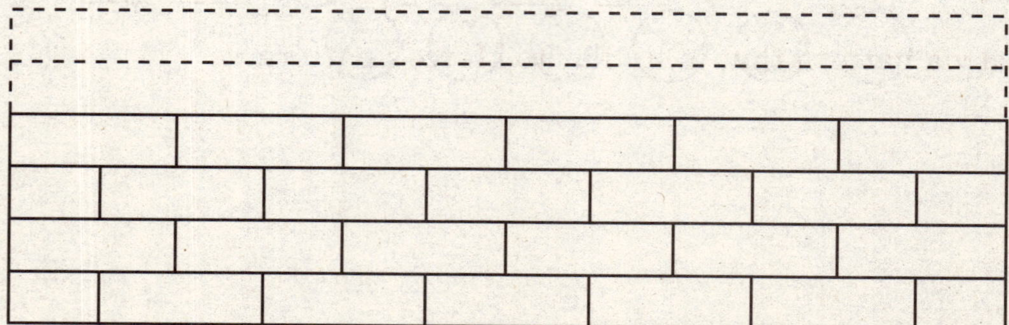

Objectives: A.A.1, A.A.3

5. Doug was given this pattern on a test and asked to draw the next figure in the pattern.

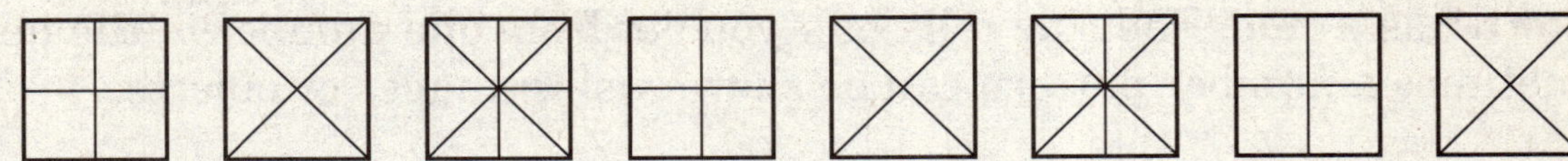

This is the figure that Doug drew.

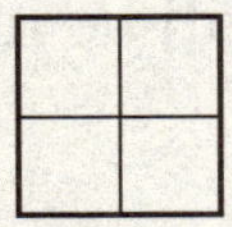

Doug's teacher told him he wasn't quite right. Fix Doug's figure to make it correct.

6. Draw your own shape pattern. Include shadings or markings in your pattern.

7. Marco drew the pattern below.

Draw a pattern that is similar to Marco's pattern.

Number Patterns

A **number pattern** is made up of different groups of numbers. A number pattern has a rule. The rule tells how you get from one number to the next. Sometimes a number pattern can be shown using shapes or objects.

Example

How many soccer balls will be in the next group in the following pattern?

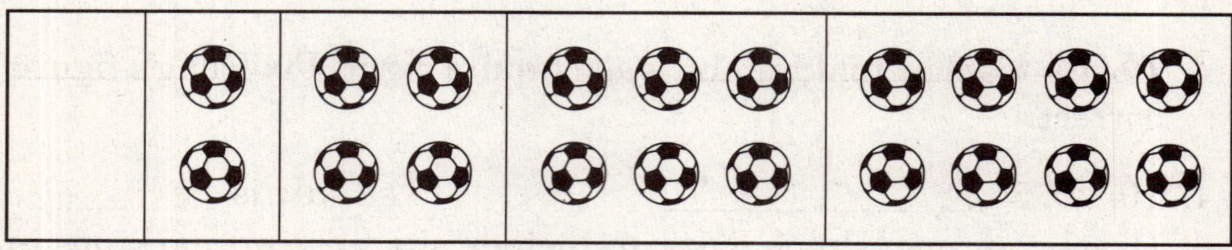

The number of soccer balls in each group follows the number pattern 0, 2, 4, 6, 8. The rule for the pattern is **+2**.

There will be 10 soccer balls in the next group. (8 **+ 2** = 10)

Example

What is the rule for the following number pattern?

24, 21, 18, 15, 12

Each number is three less than the number before it. The rule for the pattern is **−3**.

To make a similar number pattern, follow the same rule, except start with a different number.

Example

Make a pattern similar to the pattern in the example above.

The rule for the number pattern above is −3. Start with a different number and use the rule −3.

16, 13, 10, 7, 4

Objectives: A.A.1, A.A.3

Practice

Directions: For Numbers 1 through 7, write the next two numbers in the pattern. Then write the rule for the pattern.

1. 6, 10, 14, ______, ______ The rule is ________ .

2. 40, 32, 24, ______, ______ The rule is ________ .

3. 15, 13, 11, ______, ______ The rule is ________ .

4. 10, 30, 50, ______, ______ The rule is ________ .

5. 15, 20, 25, ______, ______ The rule is ________ .

6. 36, 30, 24, ______, ______ The rule is ________ .

7. 1, 4, 7, ______, ______ The rule is ________ .

Directions: For Numbers 8 and 9, make a pattern similar to the given pattern.

8. 2, 4, 6, 8, 10 ________________________

9. 88, 78, 68, 58, 48 ________________________

10. Circle the group that **does not** follow a pattern.

    5, 10, 15, 20, 25 3, 5, 7, 9, 12 5, 4, 3, 2, 1

    What could you change in the circled group to make it a pattern?

    ____

Number Patterns in Tables

Number patterns can be used to solve problems. You can make a table that shows the pattern to help you solve the problem.

Example

A folding chair has 4 legs. How many legs are there on 6 chairs?

To solve this problem, make a table.

Chairs	1	2	3	4	5	6
Legs	4	8	12	16	20	24

There are 24 legs on 6 chairs.

Practice

1. Continue the pattern from the table in the example above to find the number of legs on 8 chairs.

   There are ________ legs on 8 chairs.

2. Jamie will put pennies into her piggy bank each day this week. The first day she put 1 penny into her piggy bank. Each day she will double the number of pennies that she put in the day before. How many pennies will Jamie put into her piggy bank on the seventh day? Complete the pattern in the following table to answer the question.

Day	1	2	3	4	5	6	7
Number of Pennies	1	2	4	8	16		

   Jamie will put ________ pennies into her piggy bank on the seventh day.

Objectives: A.A.1, A.A.2

3. Nina is studying animals. She sees that different animals have different numbers of legs. She wrote the total number of legs for four animals in the following table. Complete the number pattern for each animal in the table.

How Many Legs?

	Bird	Mosquito	Map Turtle	Centipede
on 1	2	6	4	100
on 2	4	12	8	200
on 3				
on 4				
on 5				
on 6				

Directions: Continue the patterns from the table in Number 3 to answer Numbers 4 and 5.

4. How many map turtles would it take for there to be 36 legs?

It would take _______ map turtles.

5. How many legs are there on 10 mosquitoes?

There are _______ legs on 10 mosquitoes.

Mathematics Practice

1. Which pattern follows the rule **+5**?

   Ⓐ 11, 16, 20, 25

   Ⓑ 8, 13, 18, 23

   Ⓒ 5, 10, 20, 25

   Ⓓ 3, 5, 10, 15

2. Which figure will be next in this pattern?

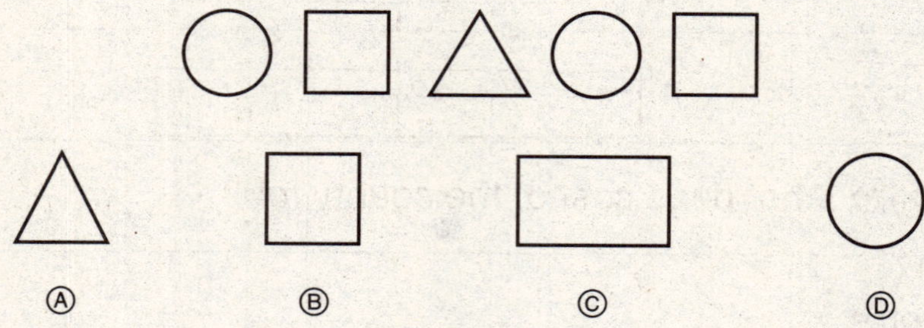

3. Which set of numbers completes the pattern shown below?

   20, 18, ____, ____, ____, 10

   Ⓐ 19, 17, 16

   Ⓑ 17, 15, 13

   Ⓒ 16, 14, 11

   Ⓓ 16, 14, 12

4. Hot dogs at the county fair are $3.00 each. The following table shows the cost for different numbers of hot dogs.

Number of Hot Dogs	Cost ($)
1	3.00
2	6.00
3	9.00
4	12.00
5	15.00
6	?
7	?

How much do 7 hot dogs cost at the county fair?

Ⓐ $7.00
Ⓑ $18.00
Ⓒ $21.00
Ⓓ $24.00

5. Which pattern is similar to the pattern shown below?

9, 18, 27, 36, 45

Ⓐ 10, 17, 26, 35, 44
Ⓑ 1, 10, 19, 28, 37
Ⓒ 81, 72, 63, 54, 45
Ⓓ 8, 19, 30, 41, 52

Lesson 10: Number Sentences

In this lesson, you will review expressions and number sentences. You will write and solve open number sentences. You will also describe change.

Expressions

An **expression** can be one of two things. It can be a single number by itself or two or more numbers using signs such as + or −. The **value** of an expression is the number the expression stands for.

Example

The following are some examples of expressions and their values.

expression:	5	$9 + 4$	$15 - 9$	$7 - 0$	$3 + 7 + 2$
	↓	↓	↓	↓	↓
value:	5	13	6	7	12

Number Sentences

A **number sentence** (also called an **equation**) shows that two expressions have equal values. The expressions are joined by an **equals sign (=)**.

Example

The following are some examples of number sentences.

$$2 + 7 = 9 \quad\rightarrow\quad 9 = 9$$

$$12 - 5 = 1 + 6 \quad\rightarrow\quad 7 = 7$$

$$14 - 1 = 2 + 4 + 7 \quad\rightarrow\quad 13 = 13$$

TIP: The expressions on either side of the equals sign can be switched around.

Objectives: A.B.1, A.B.3

Finding Missing Numbers

Sometimes number sentences may be missing a number. Symbols such as ○, □, and △ are used to stand for missing numbers. These symbols can stand for different numbers in different problems. Number sentences that are missing a number are sometimes called **open number sentences**.

Examples

The following are two examples of open number sentences.

□ + 7 = 12

13 − △ = 4

To find a missing number in an open number sentence, use addition or subtraction facts. There is only one number that makes this type of open number sentence true.

Example

What number does the □ stand for in the following open number sentence?

□ + 7 = 12

Use addition facts to find the number you add to 7 to equal 12.

5 + 7 = 12

The □ stands for the number 5. This is written □ = 5.

Example

What number does the △ stand for in the following open number sentence?

13 − △ = 4

Use subtraction facts to find the number you must subtract from 13 to equal 4.

13 − **9** = 4

The △ stands for the number 9. This is written △ = 9.

Practice

Directions: For Numbers 1 through 10, find the number that the symbol stands for in each open number sentence.

1. ○ + 7 = 16 ○ = ________
2. 13 − □ = 5 □ = ________
3. △ + 2 = 2 △ = ________
4. 10 − ○ = 6 ○ = ________
5. 1 + ○ = 3 ○ = ________
6. 0 + △ = 4 △ = ________
7. 11 − □ = 7 □ = ________
8. 12 − ○ = 9 ○ = ________
9. □ + 8 = 14 □ = ________
10. 14 − ○ = 7 ○ = ________

11. What number does the □ stand for in the following open number sentence?

    □ + 5 = 9

    A. □ = 2
    B. □ = 3
    C. □ = 4
    D. □ = 5

12. What number does the ○ stand for in the following open number sentence?

    17 − ○ = 8

    A. ○ = 7
    B. ○ = 8
    C. ○ = 9
    D. ○ = 11

Objectives: A.B.1, A.B.2, A.B.3

Writing and Solving Open Number Sentences

You can write open number sentences to solve problems. Sometimes a picture may help you write an open number sentence.

Example

Katie had 2 beads. She bought some more beads at the store. Now she has 5 beads. How many more beads did Katie buy?

The following picture shows this problem.

You can write the following number sentence to solve this problem. The ○ stands for the number of beads Katie bought at the store.

2 + ○ = 5

You can find the missing number by counting on to the number of beads Katie had before she went to the store. Count on until you get to the number of beads she has now.

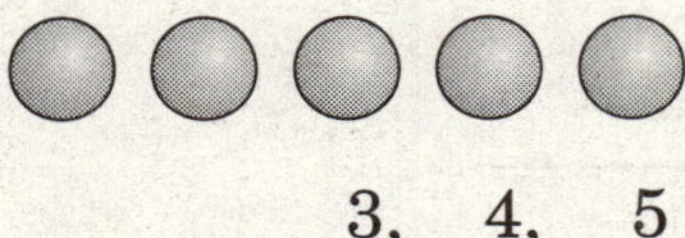

3, 4, 5

This means she bought 3 beads. This is the missing number.

2 + 3 = 5

Katie bought 3 more beads at the store.

Example

Jacob had 6 mints. He gave some away and now he has 2. How many mints did Jacob give away?

The following picture shows this problem.

You can write the following number sentence to solve this problem. The □ stands for the number of mints Jacob gave away.

6 − □ = 2

Start with the six mints that Jacob had. Cross out the mints until only 2 are left.

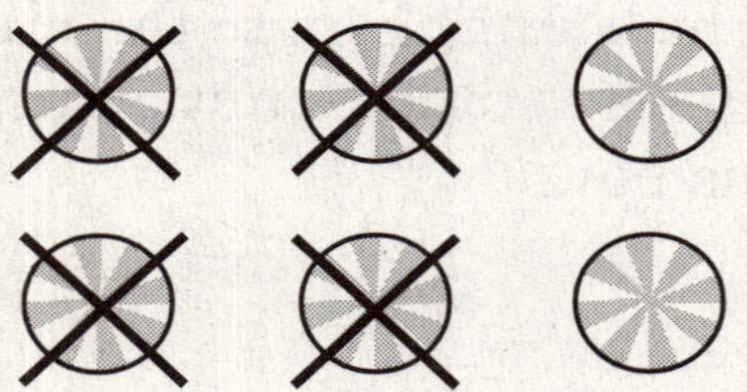

This means that Jacob gave away 4 mints, leaving 2. The missing number is 4.

6 − 4 = 2

Jacob gave away 4 mints.

Objectives: A.B.1, A.B.2, A.B.3

Practice

Directions: For Numbers 1 through 3, use the pictures to find the missing part.

1. Amy had 7 bracelets, but she lost some. Now she has 2 bracelets left. How many bracelets did Amy lose?

Amy lost _________ bracelets.

2. John's team scored 5 runs in the first inning of the kickball game. By the end of the game, his team had 12 runs. How many runs did John's team score after the first inning?

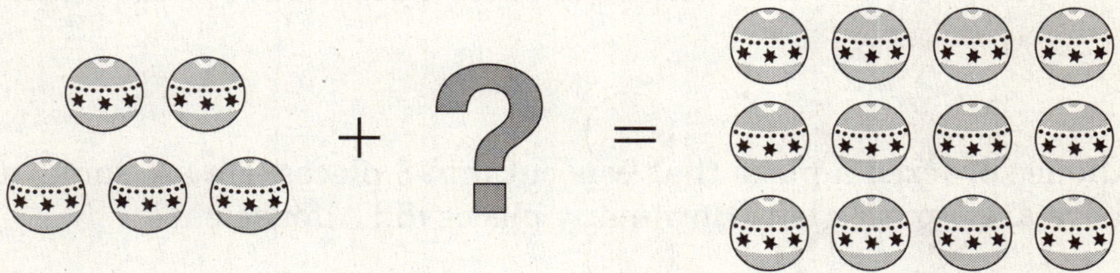

John's team scored _________ runs after the first inning.

3. Drew must eat 8 beans with his dinner. He has 3 beans left to eat. How many beans did Drew already eat?

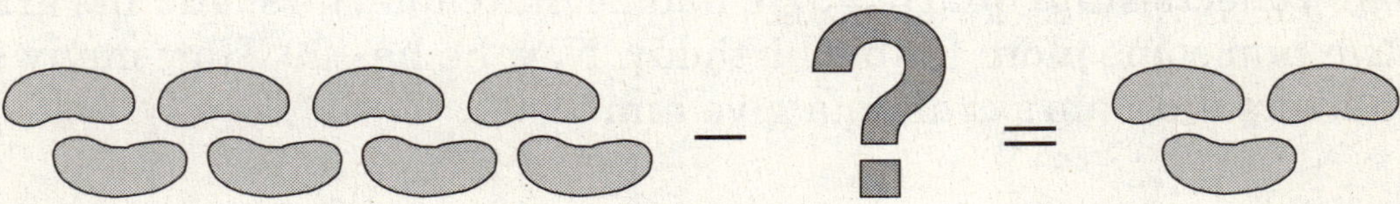

Drew already ate _________ beans.

Directions: For Numbers 4 through 7, write a number sentence to help you find the missing part.

4. Marco picked 5 oranges. Susanna also picked some oranges. Altogether, Marco and Susanna picked 16 oranges. How many oranges did Susanna pick?

5. Bree had $20 in her purse when she went to the mall. She spent some of her money on a new book. Now she has $8 left in her purse. How much did the book cost?

6. Alfonse ordered a pizza that was cut into 8 pieces. He ate some, and now there are 5 pieces left. How many pieces did Alfonse eat?

7. Sam collects state quarters. He had 24 state quarters, and his grandma gave him some more for his birthday. Now he has 29. How many state quarters did Sam's grandma give him?

Describing Changes

There are two different ways to describe a change that has taken place.

One way is to use **words** to describe the change that has taken place. Some of the words that you may use to describe a change are *fewer*, *less than*, *smaller*, *more*, *greater than*, *bigger*, *larger*, and *taller*.

The other way is to use **numbers from measuring** to describe the change. This way of describing changes is usually more exact than words.

Example

The second graders in Mr. Wang's class measured the height of a plant in centimeters each week for four weeks. Then they made the following table to show their measurements.

Plant Height

Week	Height (in centimeters)
1	1
2	3
3	8
4	14

Marie noticed from the table that the plant is growing taller each week.

Jared noticed that the plant grew 1 centimeter the first week, 2 centimeters the second week, 5 centimeters the third week, and 6 centimeters the fourth week.

Both Marie and Jared described the change in the height of the plant. Marie described the change in height using words: *growing taller*. Jared described the change in height using numbers from measuring: *1 centimeter*, *2 centimeters*, *5 centimeters*, and *6 centimeters*.

Practice

Directions: Use the following information to answer Numbers 1 and 2.

The temperature at 6:00 P.M. was 35°F. The temperature at 8:00 P.M. was 29°F.

1. Use words to describe the change in temperature.

____

____

2. Use numbers from measuring to describe the change in temperature.

____

____

Directions: Use the following information to answer Numbers 3 and 4.

Eloise had 2 quarters. Then she found a dime and a nickel.

3. Use words to describe the change in the amount of money Eloise has.

____

____

4. Use numbers from measuring to describe the change in the amount of money Eloise has.

____

____

Mathematics Practice

1. What number belongs in place of the □?

   □ − 10 = 7

   Ⓐ 3
   Ⓑ 7
   Ⓒ 13
   Ⓓ 17

2. Ed had 2 watermelons before he went to the store. After he went to the store, he had 6 watermelons. How many watermelons did Ed buy at the store?

   Which open number sentence could you use to solve the problem?

   Ⓐ 2 − △ = 6
   Ⓑ △ − 2 = 6
   Ⓒ △ + 6 = 2
   Ⓓ 2 + △ = 6

3. Kaci was 42 inches tall last March. She is now 47 inches tall. Which of the following correctly uses words to describe Kaci's height now as compared to her height in March?

   Ⓐ Kaci is taller now.
   Ⓑ Kaci is the same height.
   Ⓒ Kaci was taller in March.
   Ⓓ Kaci did not grow at all.

4. A total of 17 students are signed up to run either the 50-meter race or the 75-meter race at the school carnival. 8 students are signed up to run the 50-meter race. How many students are signed up to run the 75-meter race? Let the □ stand for the number of students signed up to run the 75-meter race.

   Which open number sentence could you use to solve the problem?

   Ⓐ □ − 17 = 8

   Ⓑ □ + 17 = 8

   Ⓒ □ + 8 = 17

   Ⓓ 17 + 8 = □

5. Jonah had 8 friends over to his house after school. Some of the friends have gone home. There are still 2 friends at Jonah's house. How many friends have gone home?

   Ⓐ 5

   Ⓑ 6

   Ⓒ 7

   Ⓓ 8

6. What number does the ○ stand for in the following open number sentence?

   3 + ○ = 6

   Ⓐ 2

   Ⓑ 3

   Ⓒ 4

   Ⓓ 5

Unit 3

Geometry

Look around you. Everywhere you look, you see shapes and solids. Whether it's a rectangular football field or a town square, there are circles, triangles, squares, rectangles, prisms, cylinders, spheres, and pyramids all around you.

In this unit, you will review plane figures and solid figures and how they look. You will put together and take apart figures. You will learn what plane and solid figures look like from different positions. You will also learn about symmetry, congruent and similar figures, shapes in motion, and the coordinate grid.

In This Unit

Lesson 11: Geometric Shapes

In this lesson, you will review plane and solid figures.

Plane Figures

Plane figures are flat, and they have length and width. Most plane figures have **sides (edges)** and **corners (angles)**. Other figures have **curves**. Here are some plane figures:

circle — A **circle** is a round, curved figure with no straight sides and no corners.

triangle (side, corner) — A **triangle** has 3 sides and 3 corners.

square — A **square** has 4 equal sides and 4 square corners.

rhombus — A **rhombus** has 4 equal sides and 4 corners.

rectangle — A **rectangle** has 4 sides and 4 square corners.

trapezoid — A **trapezoid** has 4 sides, 4 corners, and 1 pair of parallel sides.

hexagon — A **hexagon** has 6 sides and 6 corners.

parallel lines / parallel sides — **Parallel lines/parallel sides** are lines/sides that never come together or cross even if you continue them forever.

A **quadrilateral** is any four-sided plane figure. Squares, rhombuses, rectangles, and trapezoids are quadrilaterals. A **parallelogram** is a quadrilateral that has two pairs of parallel sides. Squares, rhombuses, and rectangles are parallelograms.

Objectives: G.A.1, G.A.2

Practice

1. Count the sides and angles of each plane figure.

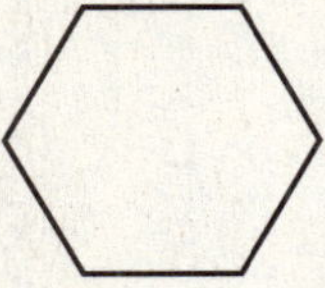

sides ________

angles ________

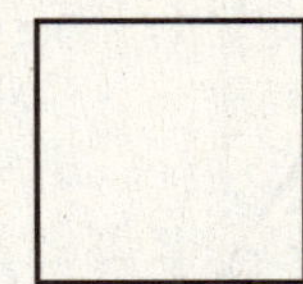

sides ________

angles ________

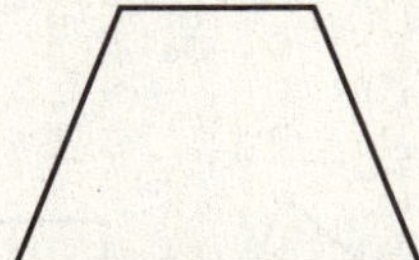

sides ________

angles ________

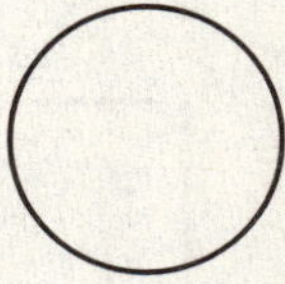

sides ________

angles ________

2. Draw a line from each real-life object to the plane figure it most looks like.

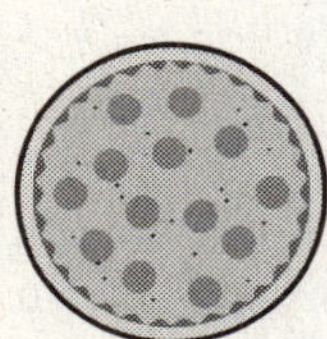

rectangle circle triangle square

3. Color the plane figure that has exactly one pair of parallel sides.

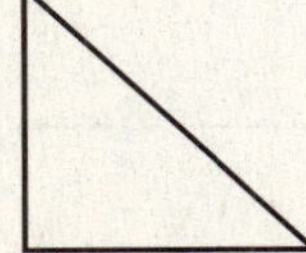

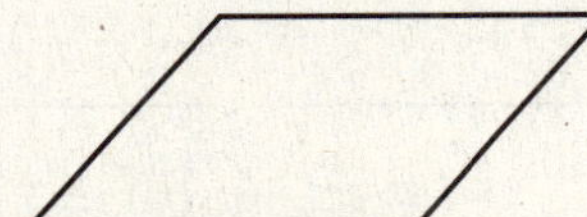

Directions: For Numbers 4 through 7, read the sentence. Then write the letter of each figure that matches the sentence. You may have more than one answer.

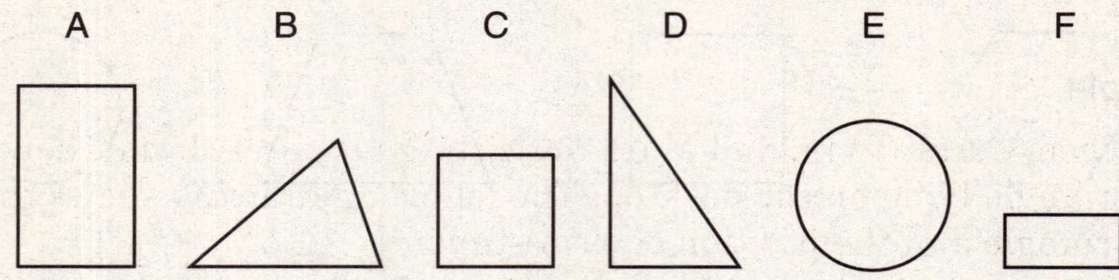

4. It has no sides or angles. ____________

5. It is a triangle. ____________

6. It has four sides and four angles. ____________

7. It is a square. ____________

8. Name one way a square and a rectangle are alike and one way they are different.

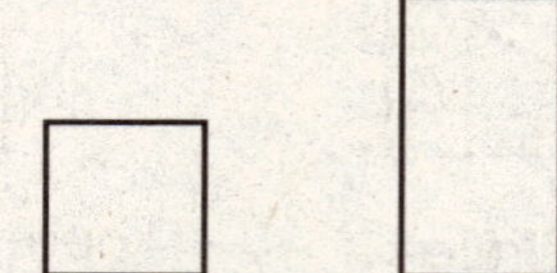

Alike ____

____

Different ____

____

Objectives: G.D.1

Plane Figures from Different Positions

Some plane figures look different when they are seen from different positions. The shape of the plane figures will remain the same.

Example

Theo, Coral, Tami, and Mark are sitting around a desk. There is a triangle lying on the desktop. The following drawing shows the triangle and the position of each student.

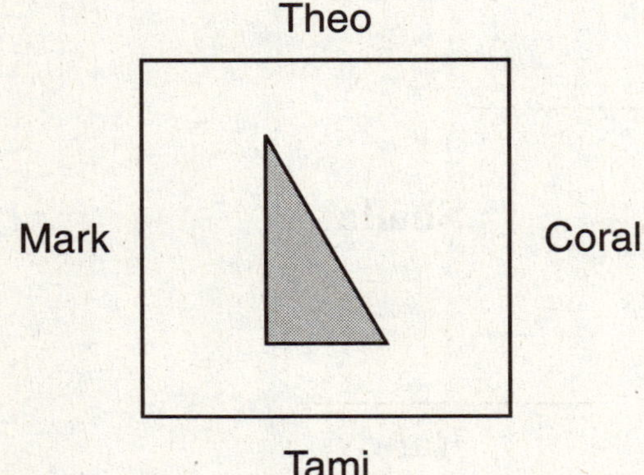

The following drawings show what the triangle looks like to each student.

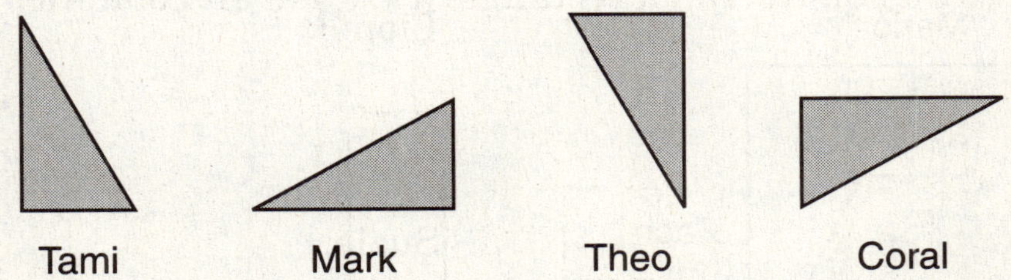

Each student sees the same triangle in a different way.

Practice

Directions: For Numbers 1 and 2, draw what the plane figure looks like to each student sitting around a desk.

1.

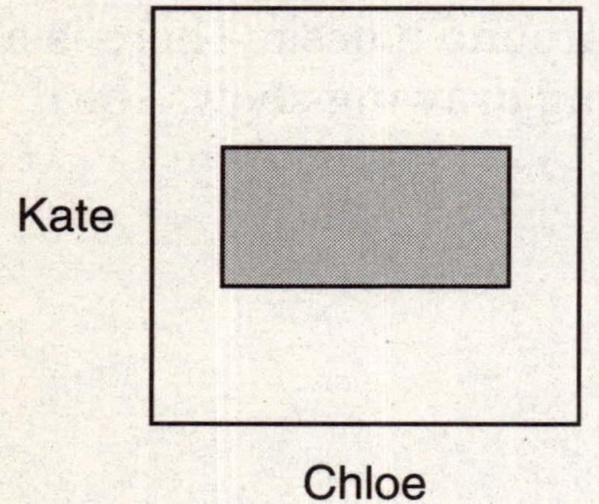

Chloe:

Kate:

Sheila:

Randy:

2.

Mario

Suzie

Zach

Lionel

Lionel:

Suzie:

Mario:

Zach:

Objectives: G.D.1

3. Which drawing shows what the following rhombus looks like to Mary?

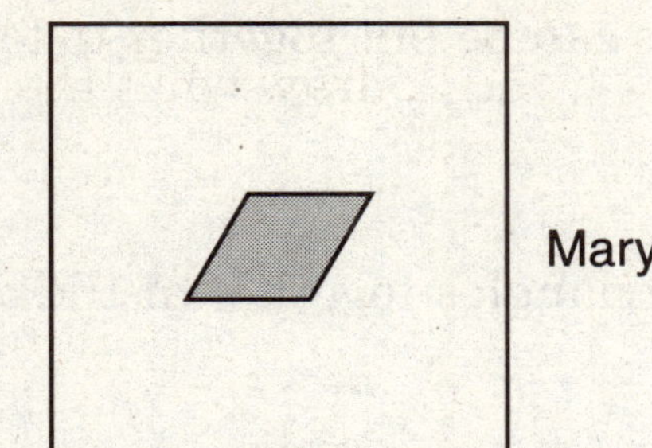

A. 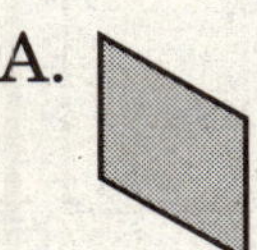

C. 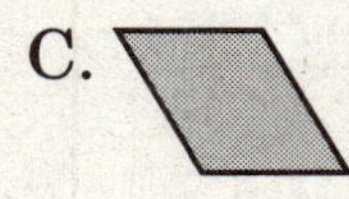

B. 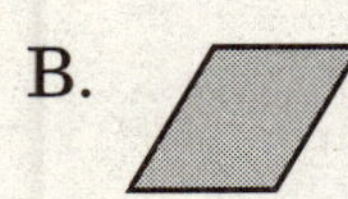

D. 

4. Which drawing shows what the following hexagon looks like to Leo?

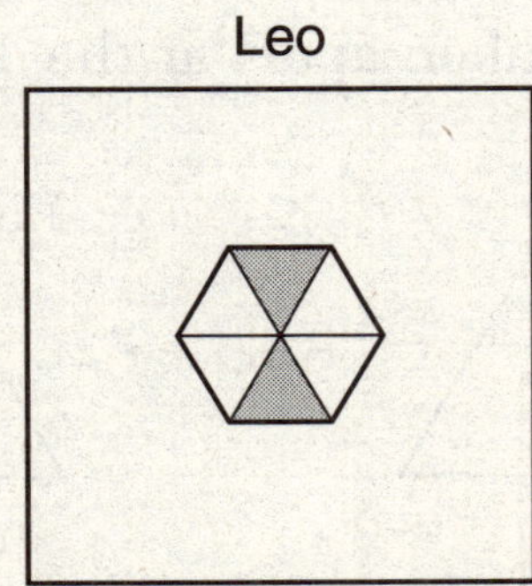

A. 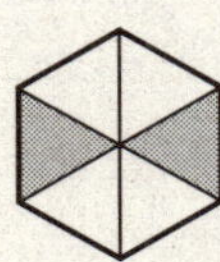

C. 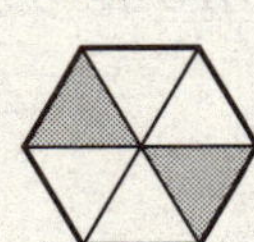

B. 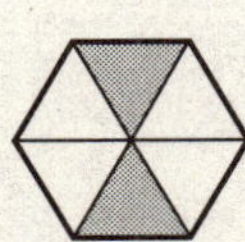

D. 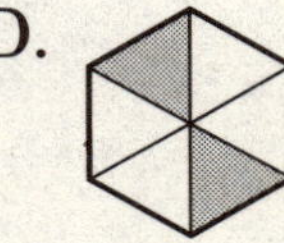

Putting Together Plane Figures

Two or more plane figures can be put together to form a single plane figure.

Example

If you put these triangles together at the dashed lines, what plane figure is formed?

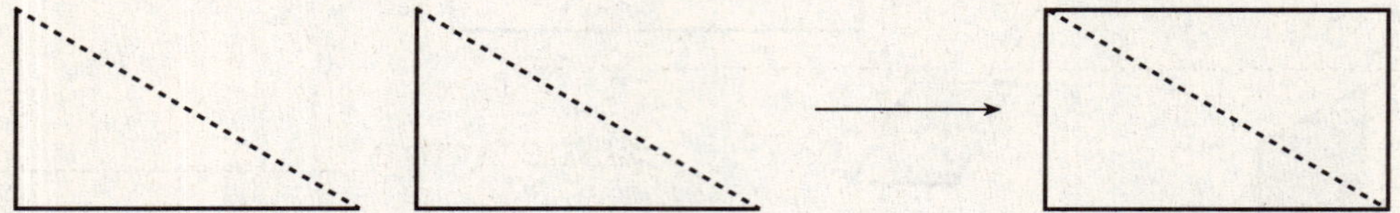

If the two triangles are put together, a rectangle is formed.

Taking Apart Plane Figures

A plane figure can be taken apart to form two or more plane figures.

Example

If the trapezoid is taken apart at the dashed line, what two plane figures are formed?

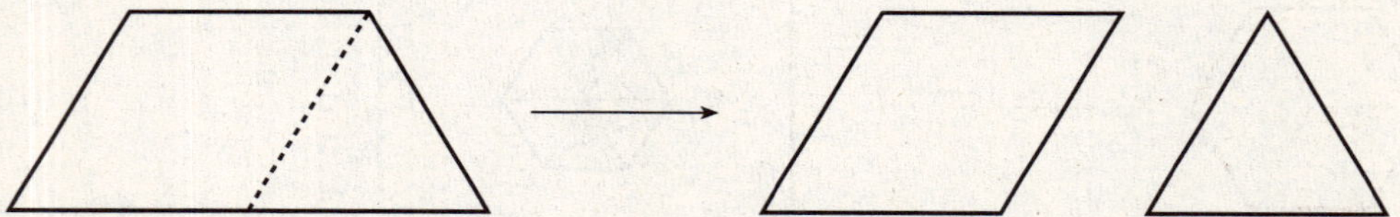

If the trapezoid is taken apart, a rhombus and a triangle are formed.

Objectives: G.A.3

Practice

Directions: For Numbers 1 and 2, draw the plane figure that is formed if the given plane figures are put together at the dashed lines. Then write the name of the plane figure.

1. 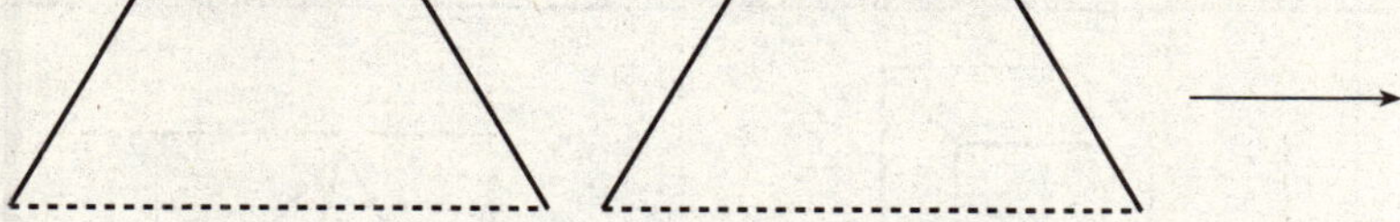

plane figure ____________________

2. 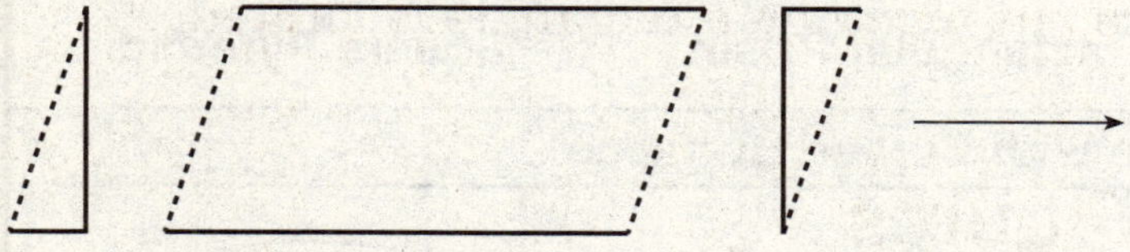

plane figure ____________________

Directions: For Numbers 3 and 4, draw the plane figures that are formed if the given plane figure is taken apart at the dashed lines. Then write the name of the plane figures.

3. 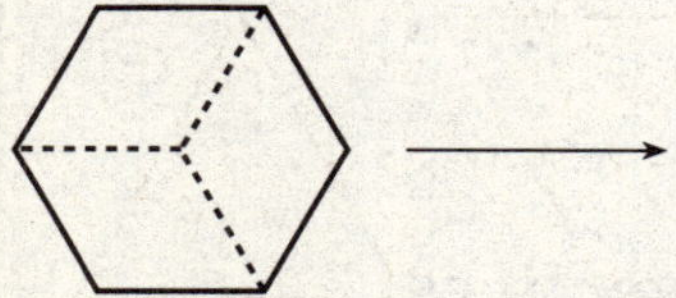

plane figures ____________________

4. 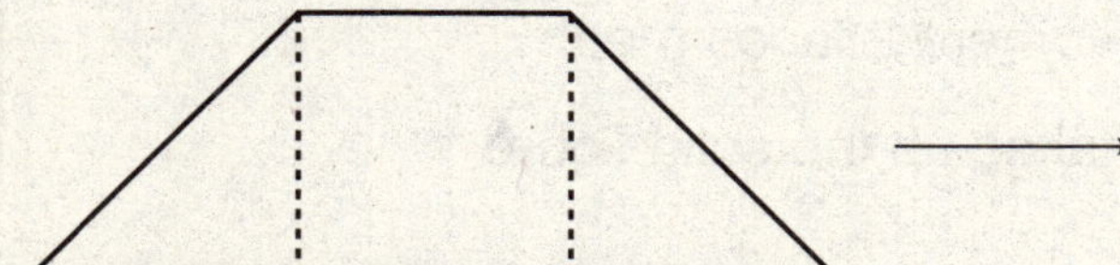

plane figures ____________________

Solid Figures

Solid figures are not flat, and they have three dimensions (length, width, and height). Some solid figures have all flat surfaces and no curves. Other solid figures have curves.

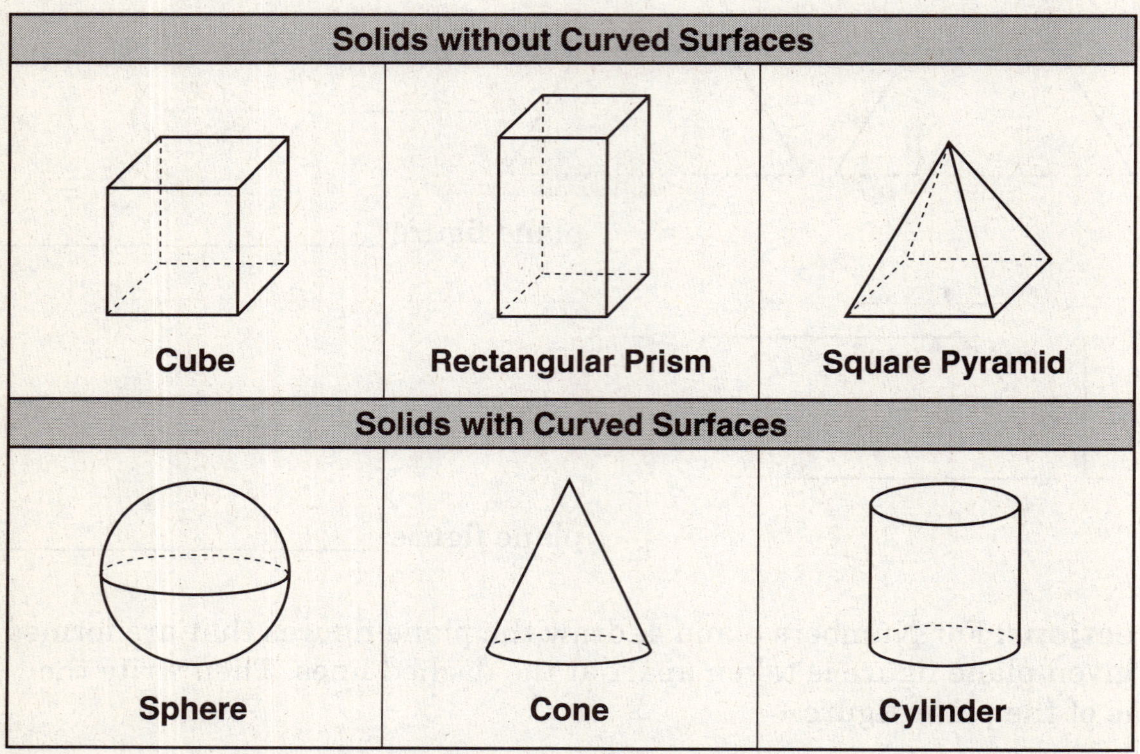

Faces, Edges, and Vertices

Solid figures without curves have **faces**, **edges**, and **vertices**.

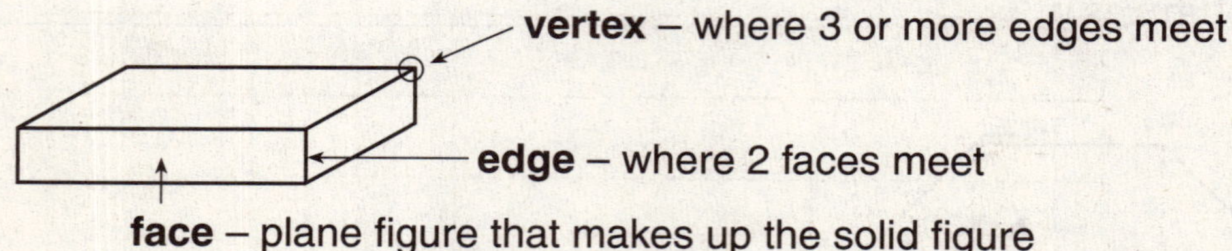

Objectives: G.A.1, G.A.2

Practice

1. Write the name of each solid figure under the object that it looks like the most.

______________ ______________ ______________

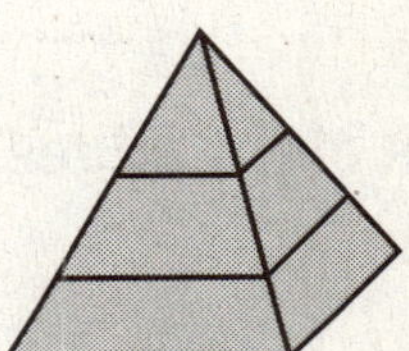

______________ ______________ ______________

2. What solid figure has no flat surfaces? ______________

3. Complete the table with the correct number of faces, edges, and vertices for each solid figure.

Solid	Faces	Edges	Vertices
Cube			
Rectangular Prism			
Square Pyramid			

Directions: For Numbers 4 through 7, read the sentence. Then write the letter of each figure that matches the sentence. You may have more than one answer.

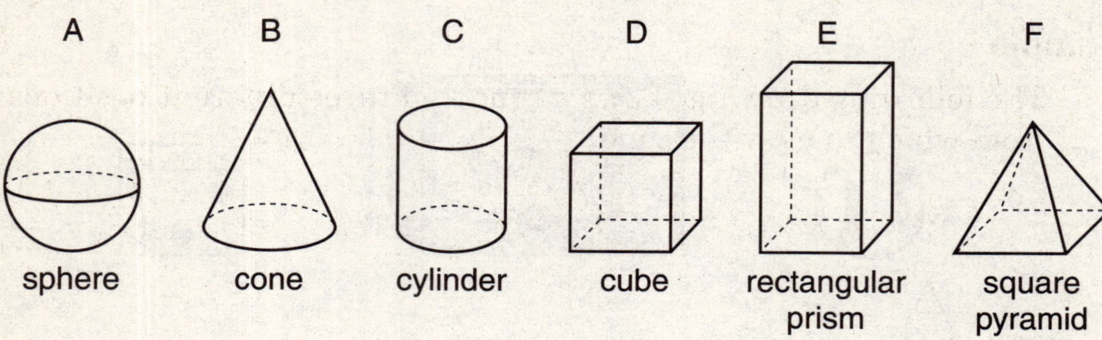

4. It has curved surfaces. ______________

5. It has eight edges. ______________

6. Some of its faces are triangles. ______________

7. It has more vertices than faces. ______________

8. How are a cube and a square pyramid **different**?

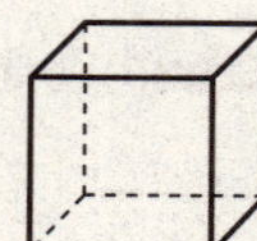 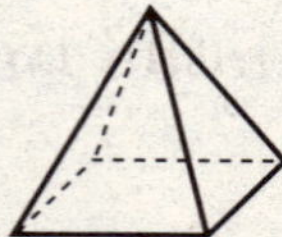

____

____

____

Objectives: G.D.1

Solid Figures from Different Positions

Some solid figures will look different when seen from different positions.

Example

The following drawing shows a cone and three different positions from which to view the cone.

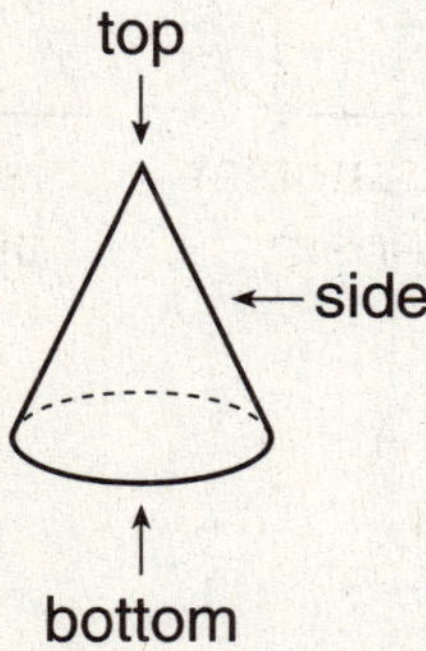

The following drawings show what the cone looks like from each position.

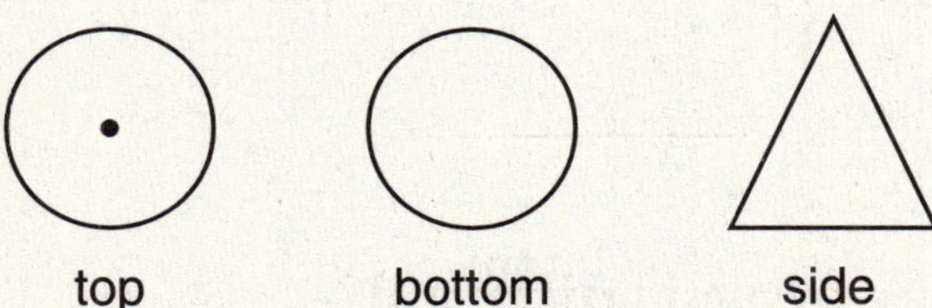

The cone looks different from the top, bottom, and side.

Practice

Directions: For Numbers 1 and 2, draw what the solid figure looks like from each position.

1.

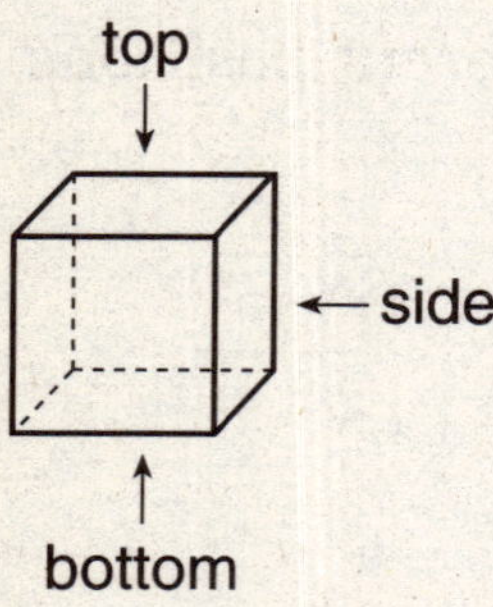

top:

bottom:

side:

2.

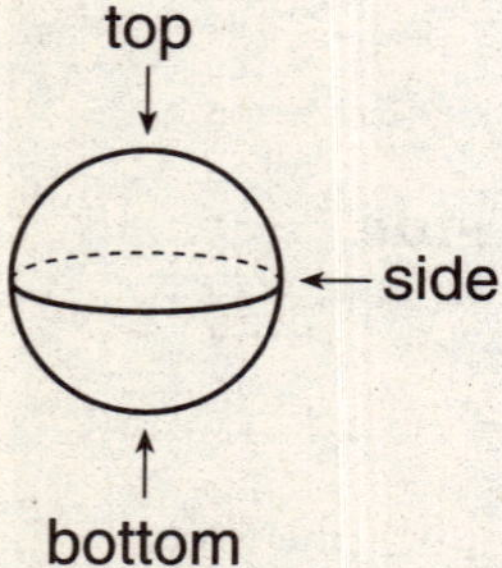

top:

bottom:

side:

Objectives: G.D.1

3. Which drawing shows what the following square pyramid looks like from the top?

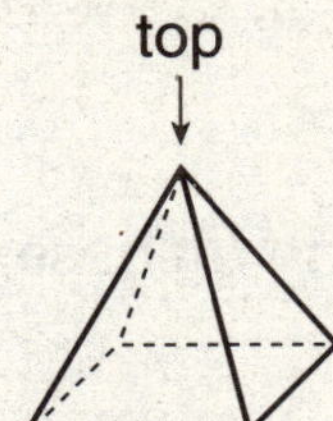

A. 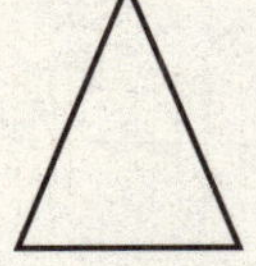

C. 

B. 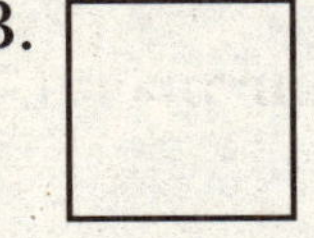

D. 

4. Which drawing shows what the following cylinder looks like from the side?

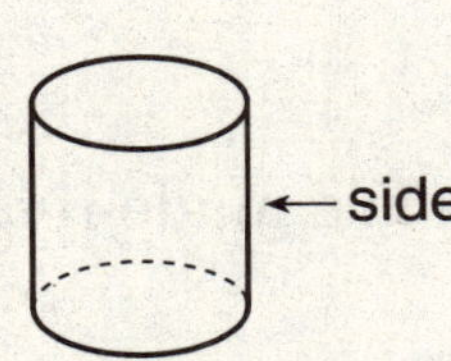

A. 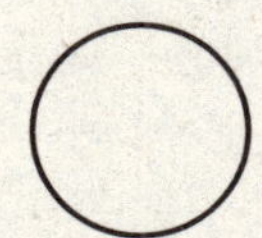

C. 

B. 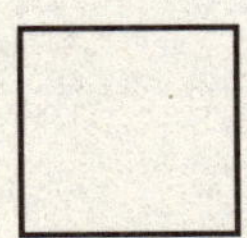

D. 

Putting Together Solid Figures

Two or more solid figures can be put together to make a single solid figure.

What solid figure is made if these cubes are put together along the shaded faces?

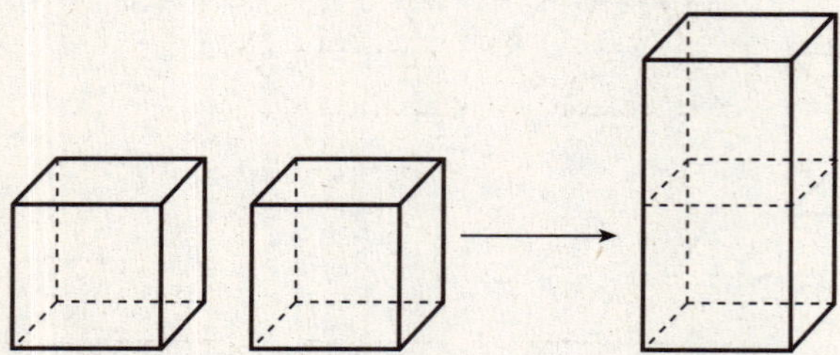

The two cubes can be put together to make a rectangular prism.

Taking Apart Solid Figures

A solid figure can be taken apart to make two or more solid figures.

What two solid figures are made if the following figure is taken apart at the shaded face?

The figure can be taken apart to make a cone and a cylinder.

Objectives: G.A.3

Practice

Directions: For Numbers 1 through 5, draw a line to match each solid figure on the left with the solid figures that make it up on the right.

1. 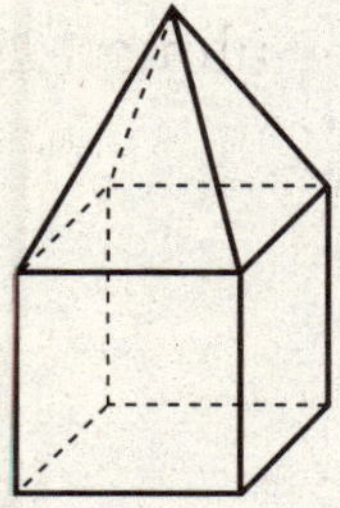

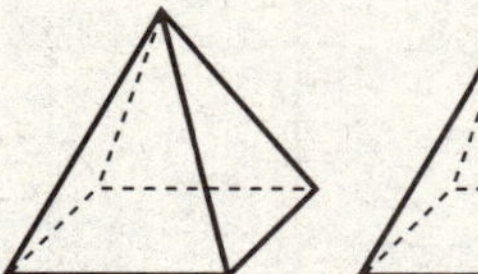

2. 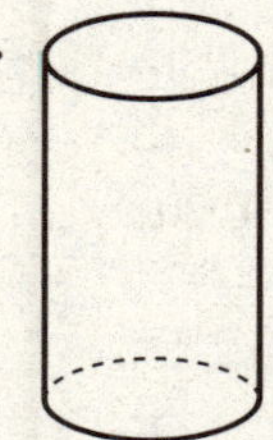

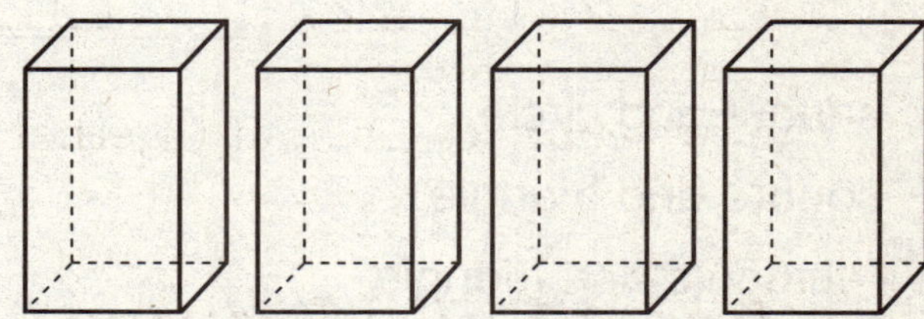

3. 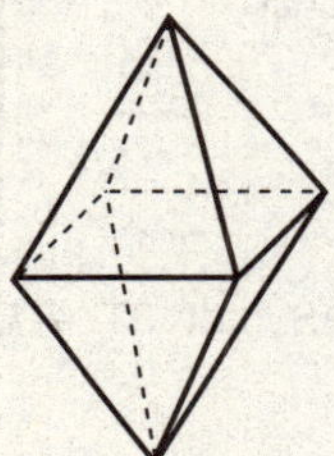

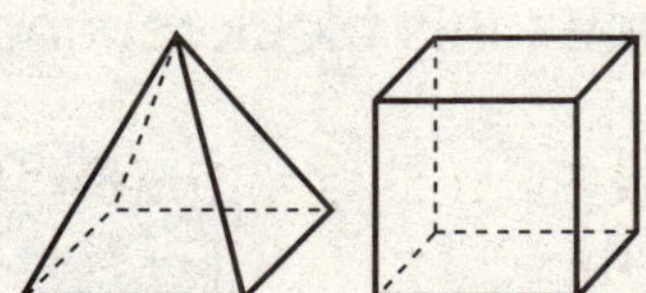

4. 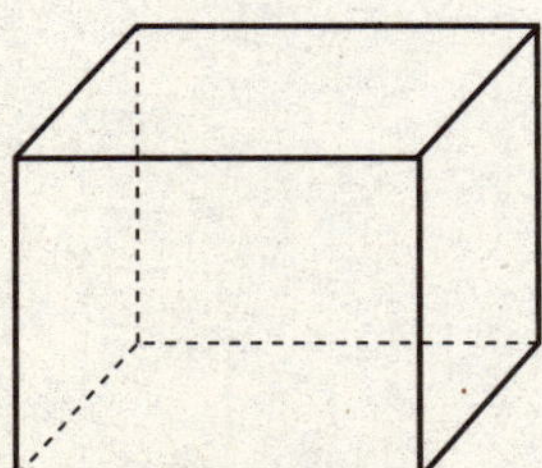

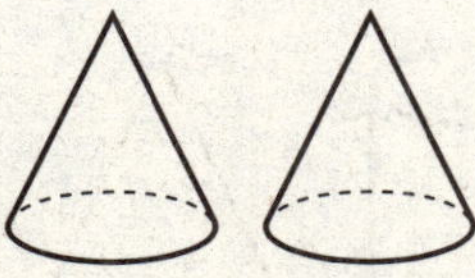

5. 

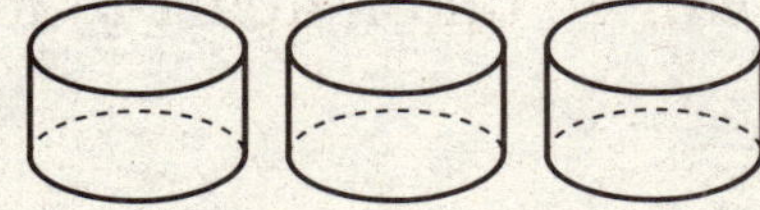

Mathematics Practice

1. What two plane figures are formed if the following figure is taken apart at the dashed line?

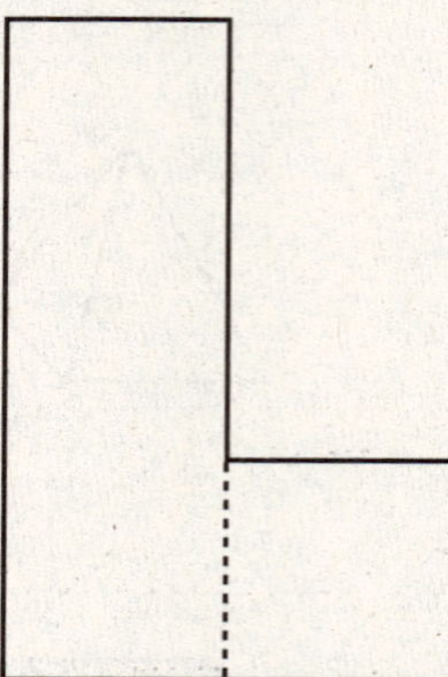

Ⓐ square and circle
Ⓑ square and triangle
Ⓒ rectangle and triangle
Ⓓ rectangle and square

2. Which solid figure does this block remind you of?

Ⓐ cone
Ⓑ cube
Ⓒ sphere
Ⓓ square pyramid

3. Which plane figure looks the same from **any** position?

 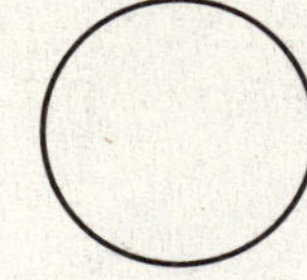

Ⓑ 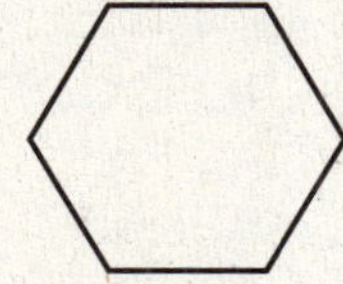

Ⓓ 

4. Which two solid figures have the same number of faces, edges, and vertices?

Ⓐ 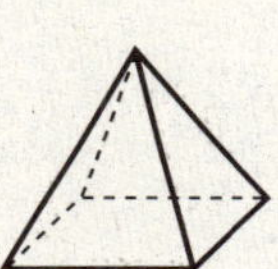 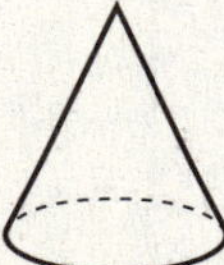

Ⓒ 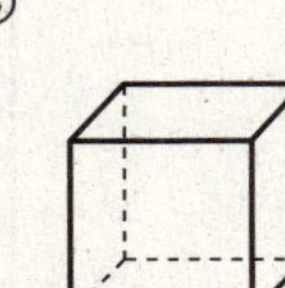 

Ⓑ 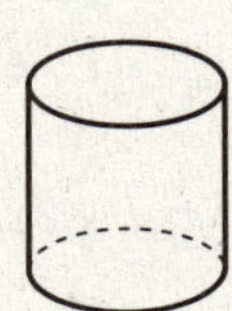 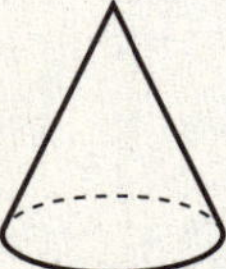

Ⓓ 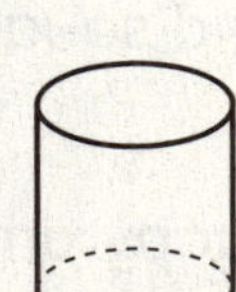 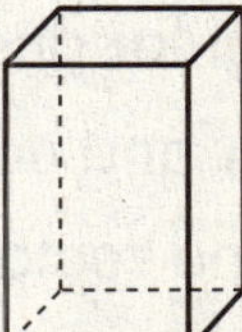

5. Which plane figure does not have 4 sides?

   Ⓐ rhombus
   Ⓑ trapezoid
   Ⓒ hexagon
   Ⓓ rectangle

6. The drawing below shows a building.

   What two solid figures are formed if the building is taken apart?

   Ⓐ cube and square pyramid
   Ⓑ rectangular prism and square pyramid
   Ⓒ cylinder and cone
   Ⓓ cone and rectangular pyramid

7. Which plane figure has no angles?

   Ⓐ triangle
   Ⓑ square
   Ⓒ trapezoid
   Ⓓ circle

Objectives: G.B.2

Lesson 12: Geometric Concepts

In this lesson, you will review symmetry, congruent and similar figures, and shapes in motion. You will also learn about the coordinate grid.

Symmetry

A figure has **line symmetry** when it can be folded so that one half of the figure fits exactly onto the other half. The line where it is folded is called the **line of symmetry**. The line of symmetry cuts the figure into two equal halves.

Example

The following figure can be folded on the dashed line so that one half fits exactly over the other.

The figure has line symmetry.

Example

The following figure cannot be folded on the dashed line or any other line so that one half fits exactly over the other.

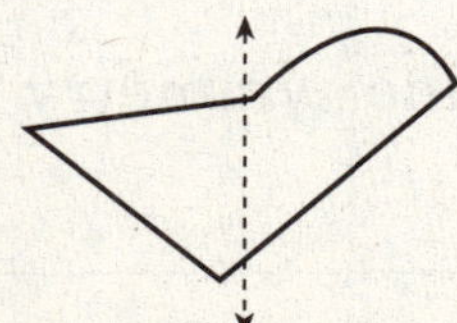

The figure does **not** have line symmetry.

Practice

1. Circle the figures below that show a line of symmetry.

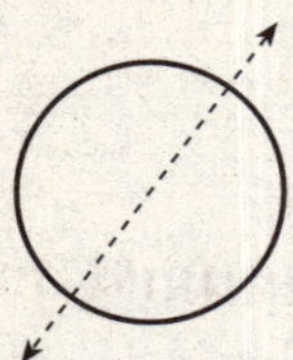
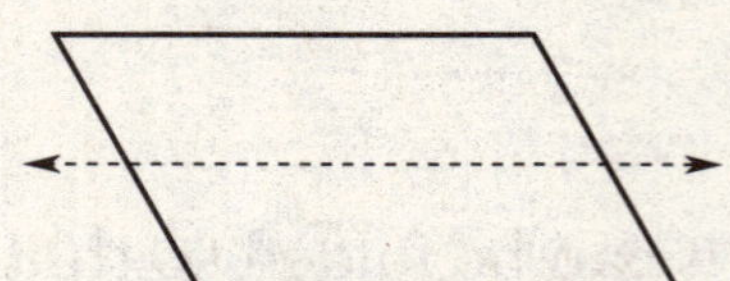
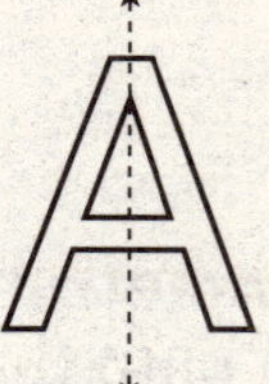
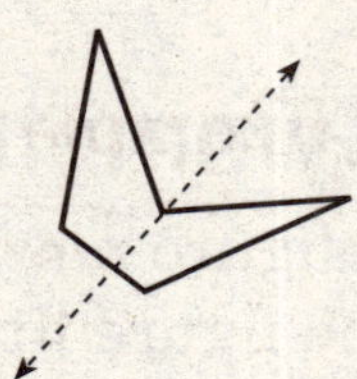

2. Draw a line of symmetry through the following square. More than one line can be drawn.

3. Each of the following figures shows half of a figure and its line of symmetry. Draw the missing half of each figure.

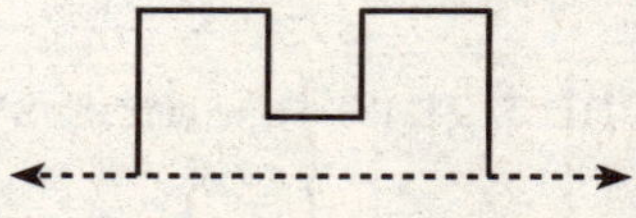

4. Write some objects in your classroom that have line symmetry.

____

____

____

____

Objectives: G.A.4

Congruent and Similar Figures

Congruent figures have the same shape and size. **Similar** figures have the same shape but do not need to have the same size. Congruent figures are also similar. But similar figures are not all congruent.

Example

The following drawing shows a pair of figures that are congruent, and a pair of figures that are similar.

The squares on the left have the same size and shape. They are best described as congruent. The triangles on the right have the same shape but have different sizes. They are similar.

Example

Tell whether the given squares are best described as congruent or similar.

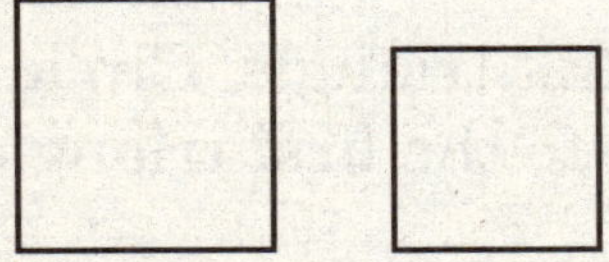

The squares are the same shape but not the same size. The squares are similar.

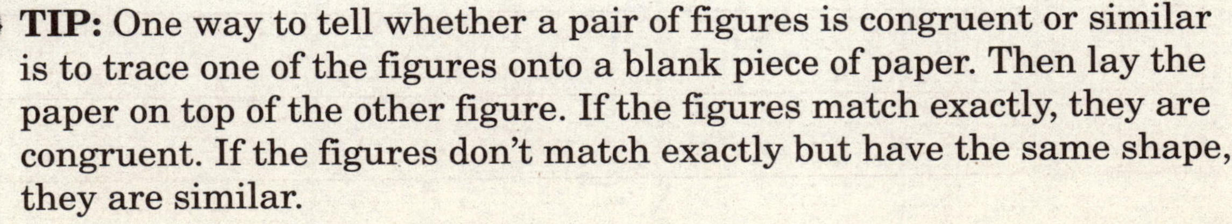

TIP: One way to tell whether a pair of figures is congruent or similar is to trace one of the figures onto a blank piece of paper. Then lay the paper on top of the other figure. If the figures match exactly, they are congruent. If the figures don't match exactly but have the same shape, they are similar.

Practice

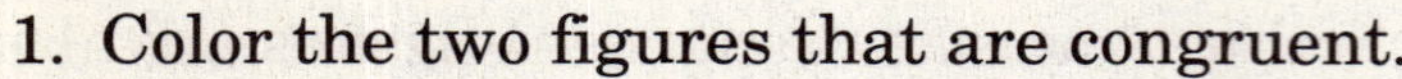

1. Color the two figures that are congruent.

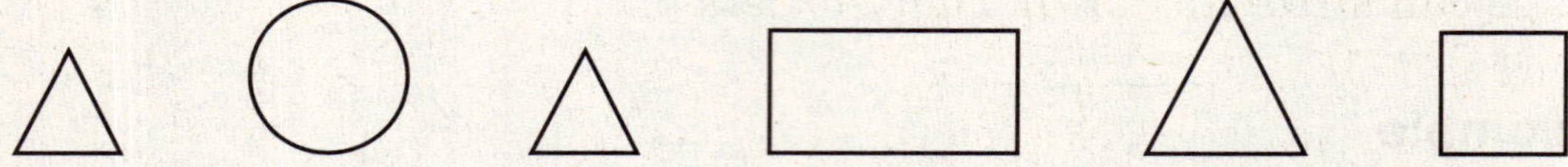

2. Color the two pairs of figures that are congruent. Use a different color for each pair. Draw lines to match the two pairs of figures that are similar but are not congruent.

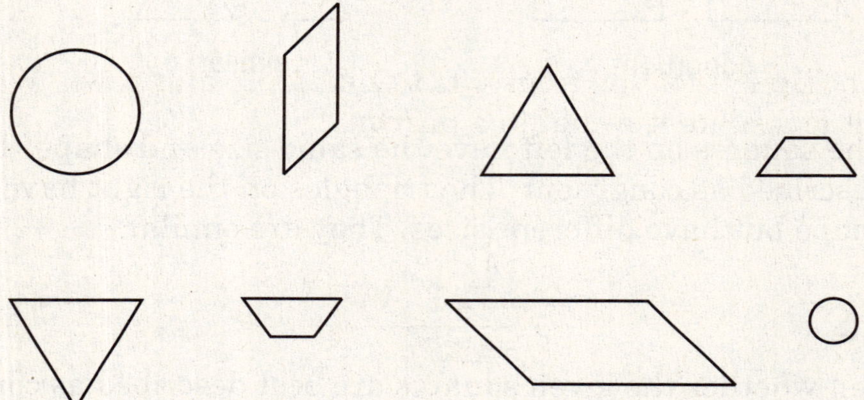

3. Color the triangle that is congruent to the first triangle. Circle the triangles that are similar but not congruent to the first triangle.

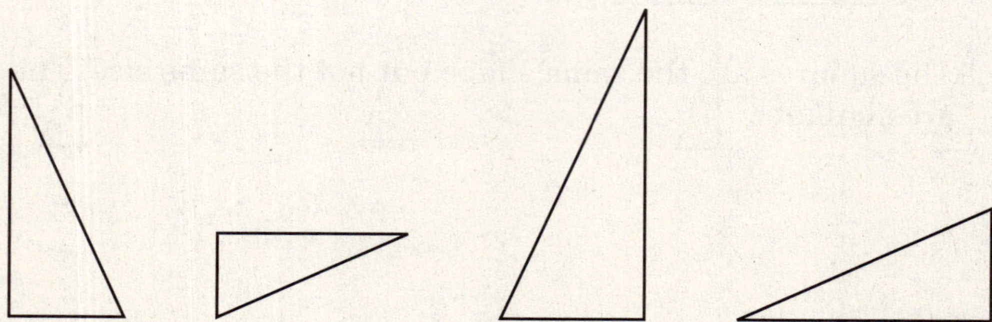

Objectives: G.B.1

Shapes in Motion

Shapes can move in different ways. They can slide, flip, or turn.

A shape can **slide** up, down, right, or left.

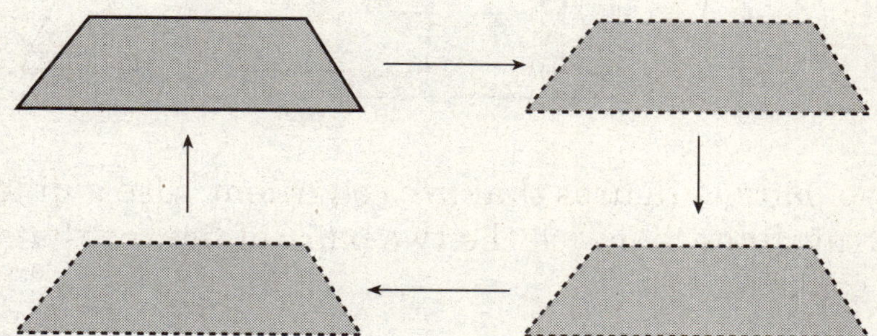

A shape can **flip** over a line from side to side or up and down. When a shape is flipped, it looks like it would in a mirror.

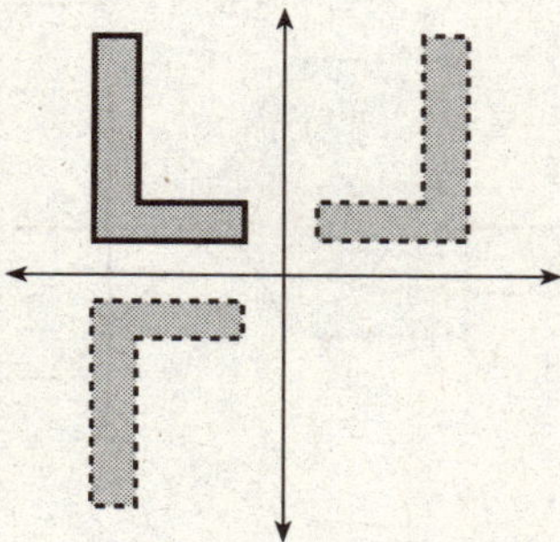

A shape can **turn** around a point. The following shows a $\frac{1}{4}$ turn, $\frac{1}{2}$ turn, $\frac{3}{4}$ turn, and 1 full turn of the triangle.

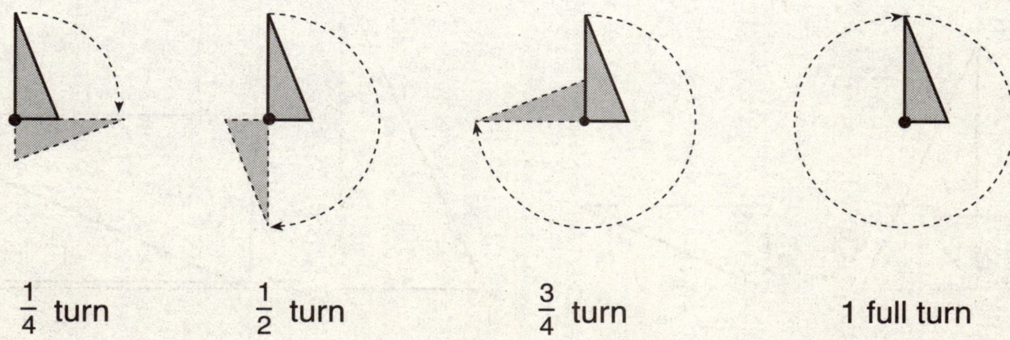

Practice

1. Draw a $\frac{1}{2}$ turn of the shape around point P.

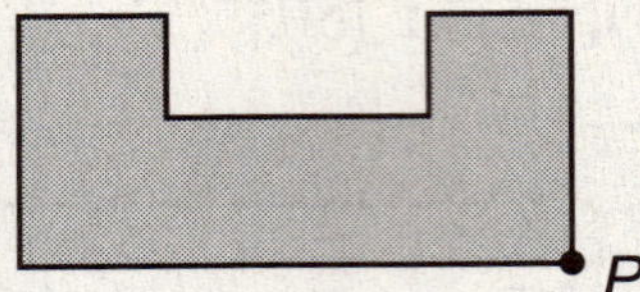

2. Draw the shape after a flip over the line.

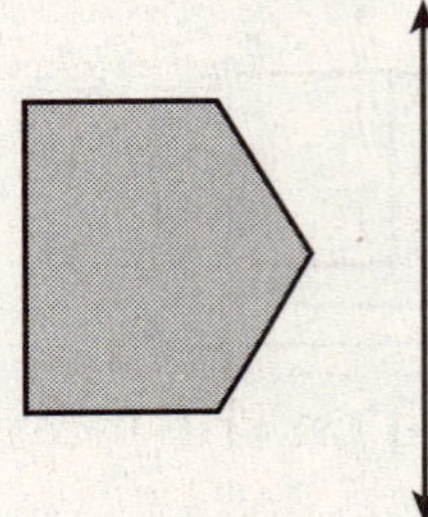

Directions: For Numbers 3 through 5, tell how each shape was moved.

3. 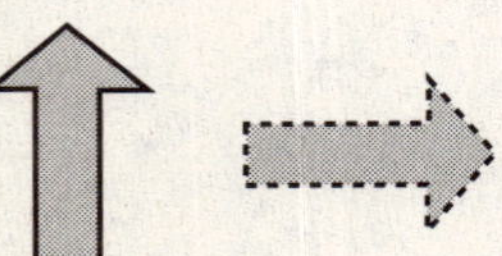

4. 

5. 

Objectives: G.C.1

The Coordinate Grid

The following grid is called a **coordinate grid**. The position of each object on a coordinate grid is described by an **ordered pair**. The first number in the ordered pair tells how far across the object is. The second number tells how far up the object is.

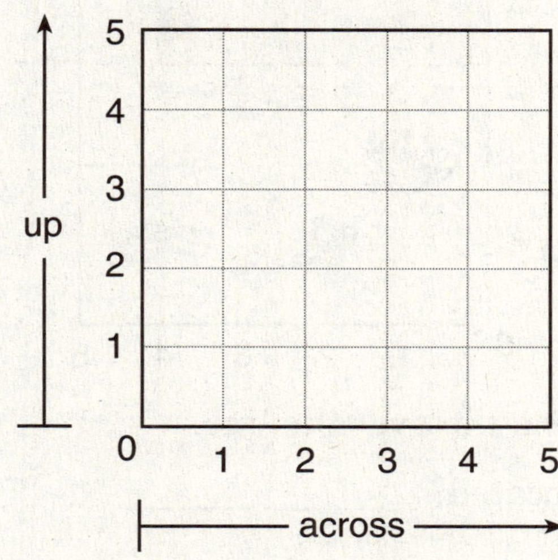

Example

Where is the football located on the following grid?

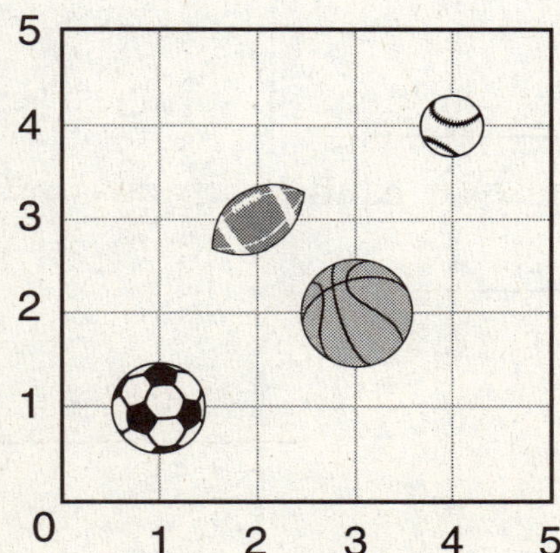

Start at 0. Move across 2. This is the first number in the ordered pair.

Move up 3. This is the second number in the ordered pair.

The football is located at **(2, 3)**.

Practice

Directions: Use the following grid to answer Numbers 1 through 8.

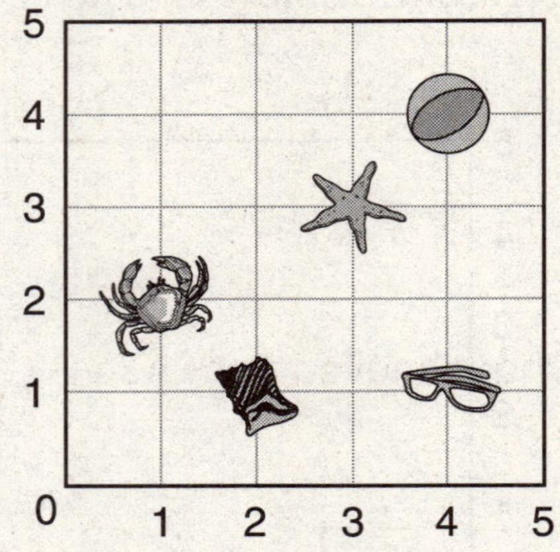

1. Where is the located? _______________

2. Where is the located? _______________

3. What object is located at (2, 1)? _______________

4. What object is located at (4, 4)? _______________

5. Where are the located? _______________

6. Draw a fish at (2, 4).

7. Draw a bucket at (3, 0).

8. Draw a shovel at (4, 3).

Mathematics Practice

1. Which heart shows a line of symmetry?

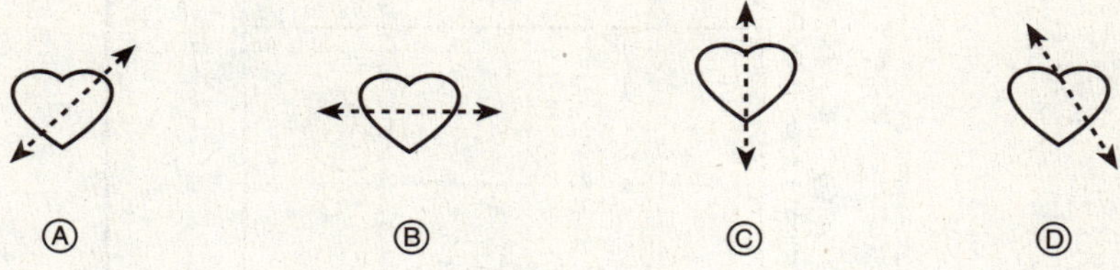

2. How was the following shape moved from A to B?

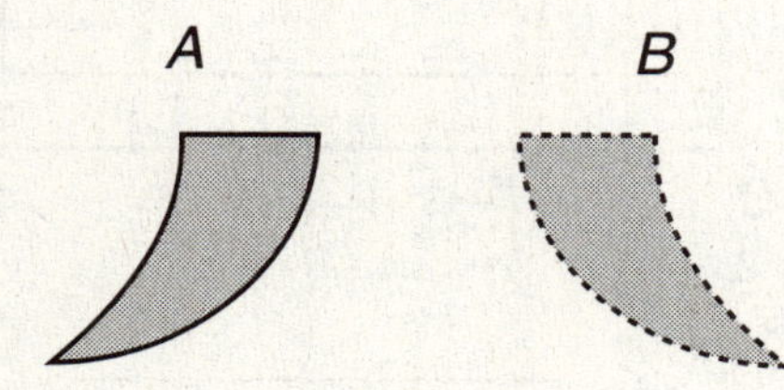

Ⓐ turn
Ⓑ flip
Ⓒ slide
Ⓓ stretch

3. Which pair of figures is congruent?

4. Where is the ☆ located on the following grid?

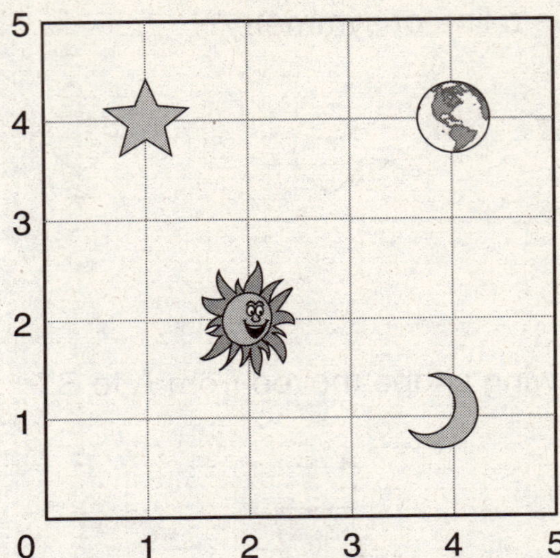

Ⓐ (4, 1)
Ⓑ (2, 2)
Ⓒ (1, 3)
Ⓓ (1, 4)

5. What does the shape below look like after a $\frac{1}{2}$ turn around the point?

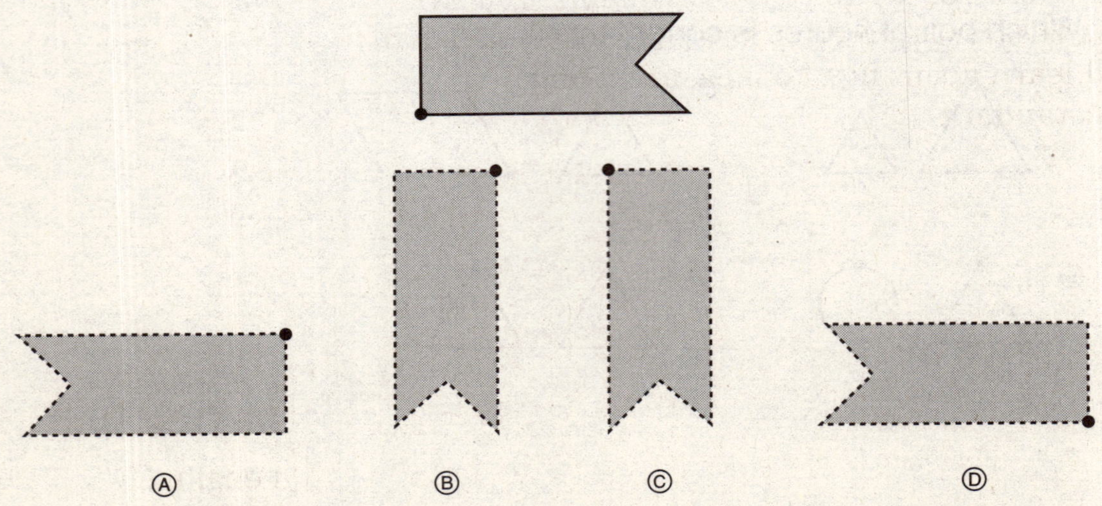

Unit 4

Measurement

Every object in the world can be measured in some way. You can measure the length of a shoe, your weight, the capacity of a glass (how much it holds), and how long you read at night.

In this unit, you will measure length, weight, and capacity using both the U.S. customary (commonly used) and metric systems. You will also tell time using two kinds of clocks, and you will learn about how to measure using nonstandard units.

In This Unit

Lesson 13: Length

In this lesson, you will learn about length. When you measure **length**, you are measuring how long, high, or far something is. You can measure length (or height or distance) using U.S. customary or metric units. Round to the nearest unit in this lesson.

U.S. Customary Units of Length

Inches (in), **feet (ft)**, and **yards (yd)** are three U.S. customary units used to measure length.

Inches

An inch is the smallest U.S. customary unit of length. Inches are used to measure short objects and distances. To measure in inches, use an inch ruler, a tape measure, or a yardstick.

Example

About how long is the following eraser in inches?

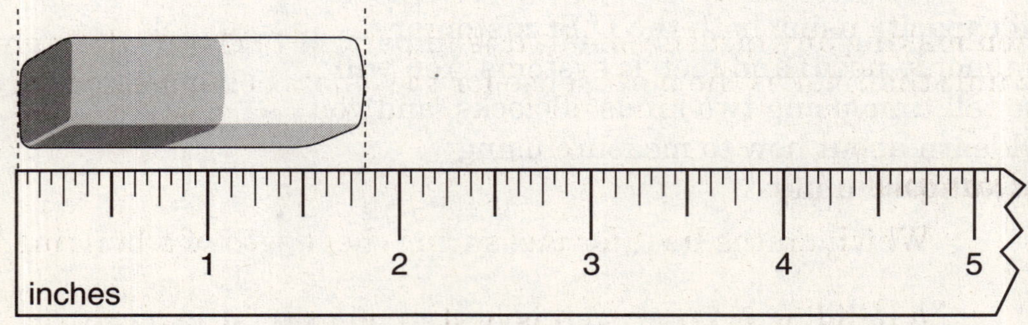

The eraser is about 2 inches long.

Objectives: M.A.1, M.A.3, M.B.2

Feet and Yards

Feet and yards are used to measure longer objects and distances. To measure in feet or yards, use an inch ruler, a tape measure, or a yardstick.

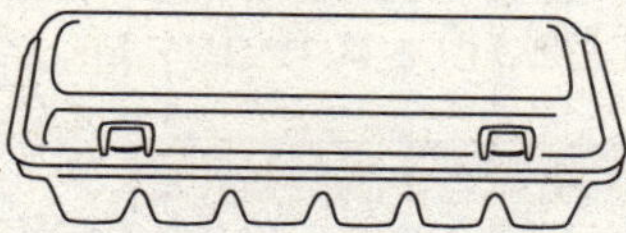

1 foot = 12 inches
A real-world egg carton is about 1 foot long.

1 yard = 3 feet = 36 inches
The distance from the top of a real-world bench to the ground is about 1 yard.

When making any measurement, it is important to choose the right unit. Pick the unit that makes the most sense for the object being measured.

Example

Which unit is **best** for measuring the length of a building?

A building is large, so it is best to measure it in yards.

Practice

Directions: For Numbers 1 through 3, first look at the object to estimate (guess) the length in inches. Then measure the length in inches.

1.

Estimate: about ______________ inches

Measurement: ______________ inches

2.

Candy Bar

Estimate: about ______________ inches

Measurement: ______________ inches

3. Your desktop

Estimate: about ______________ inches

Measurement: ______________ inches

4. Which of the items measured in Numbers 1 through 3 is the **longest**?

____

Objectives: M.A.1, M.B.2

5. Estimate the length of your teacher's desk in feet. Then measure the length in feet.

   Estimate: about ________________ feet

   Measurement: ________________ feet

6. Estimate the height of one of the walls of your classroom in yards. Then measure the height in yards.

   Estimate: about ________________ yards

   Measurement: ________________ yards

7. In the real world, which object's length would you **most likely** measure in inches? Circle the correct answer.

8. In the real world, which object's height would you **most likely** measure in feet? Circle the correct answer.

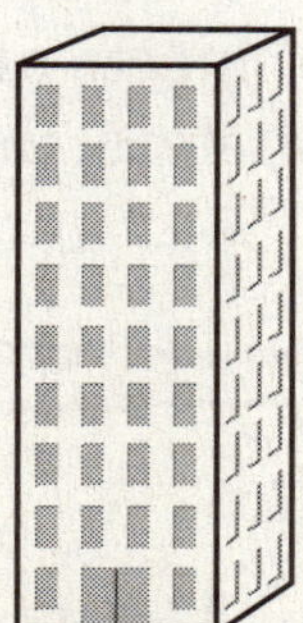

Objectives: M.A.1, M.A.3, M.B.2

Metric Units of Length

Centimeters (cm) and **meters (m)** are two metric units used to measure length.

Centimeters

Centimeters are used to measure short objects and distances. A centimeter is about half as long as an inch. To measure in centimeters, use a centimeter ruler or a meterstick.

Example

About how long is the following piece of candy in centimeters?

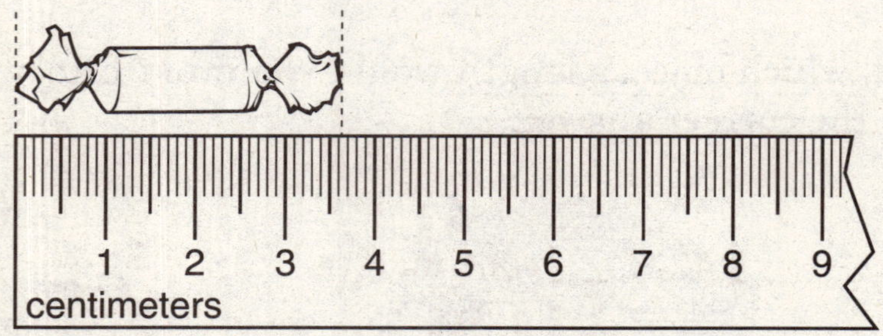

The piece of candy is about 4 centimeters long.

Meters

Meters are used to measure longer objects and distances. A meter is a little longer than a yard. To measure in meters, use a centimeter ruler or a meterstick.

1 meter = 100 centimeters
A real-world baseball bat is about 1 meter long.

Objectives: M.A.1, M.A.4, M.B.2

Practice

Directions: For Numbers 1 and 2, first look at the object to estimate the length in centimeters. Then measure the length in centimeters.

1.

Estimate: about ______________ centimeters

Measurement: ______________ centimeters

2.

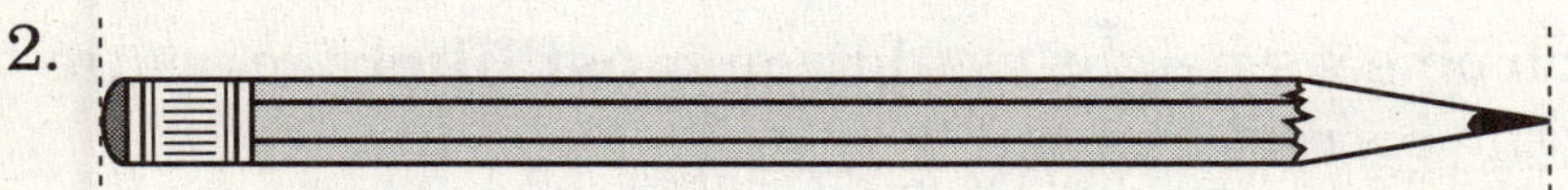

Estimate: about ______________ centimeters

Measurement: ______________ centimeters

3. Estimate the height of a bookshelf in your classroom in centimeters. Then measure the height in centimeters.

   Estimate: about ______________ centimeters

   Measurement: ______________ centimeters

4. Put the items measured in Numbers 1 through 3 in order from **shortest** to **longest**.

   ____

5. Estimate the height of your teacher in meters. Then measure his or her height in meters.

   Estimate: about _______________ meters

   Measurement: ________________ meters

6. Estimate the length of one of the walls in your classroom in meters. Then measure the length in meters.

   Estimate: about _______________ meters

   Measurement: ________________ meters

7. In the real world, which object's height would you **most likely** measure in centimeters? Circle the correct answer.

8. In the real world, which object's length would you **most likely** measure in meters? Circle the correct answer.

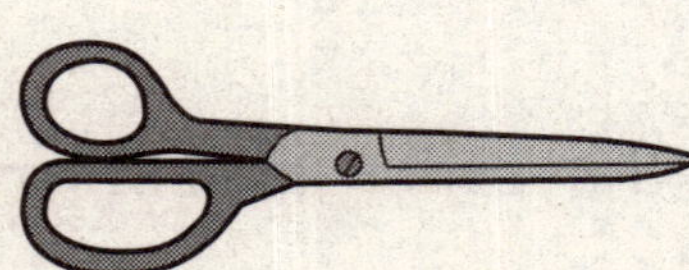

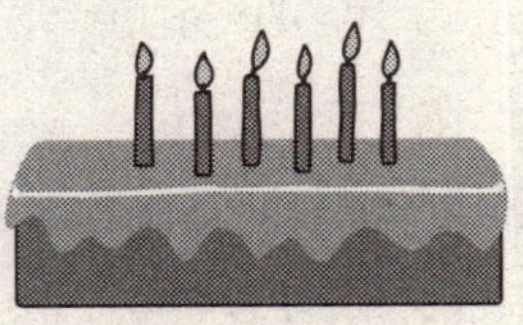

Mathematics Practice

1. Which of the following is **about** 1 foot long?

   Ⓐ a piece of chalk

   Ⓑ the top of your desk

   Ⓒ the top to the bottom of a folder

   Ⓓ the chalkboard

2. Which unit of length would you **most likely** use to measure the length of a pair of scissors?

   Ⓐ centimeter

   Ⓑ meter

   Ⓒ foot

   Ⓓ yard

3. Which of the following is **about** 1 centimeter long?

   Ⓐ the length of your foot

   Ⓑ the length of your arm

   Ⓒ the width of a piece of hair

   Ⓓ the width of your finger

4. Which length is the **longest**?

   Ⓐ 2 inches

   Ⓑ 2 yards

   Ⓒ 2 centimeters

   Ⓓ 2 feet

5. How many centimeters are in 1 meter?

   Ⓐ 12
   Ⓑ 36
   Ⓒ 100
   Ⓓ 1,000

Directions: Use the following key to answer Numbers 6 and 7. Use your ruler to measure. Round your answer to the nearest unit.

6. **About** how long is the key in inches?

   Ⓐ 2 inches
   Ⓑ 8 inches
   Ⓒ 10 inches
   Ⓓ 12 inches

7. **About** how long is the key in centimeters?

   Ⓐ 1 centimeter
   Ⓑ 5 centimeters
   Ⓒ 10 centimeters
   Ⓓ 20 centimeters

Objectives: M.A.1, M.A.3, M.B.2

Lesson 14: Weight

In this lesson, you will learn about weight. When you measure **weight**, you are measuring how heavy something is. You can measure weight using U.S. customary or metric units. Round to the nearest unit in this lesson.

U.S. Customary Units of Weight

Ounces (oz) and **pounds (lb)** are two U.S. customary units used to measure weight.

Ounces

Ounces are used to weigh light objects.

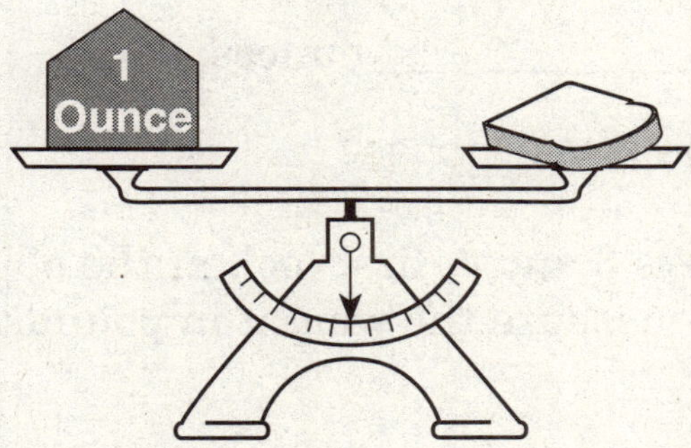

A real-world slice of bread weighs about 1 ounce.

Pounds

Pounds are used to weigh heavier objects.

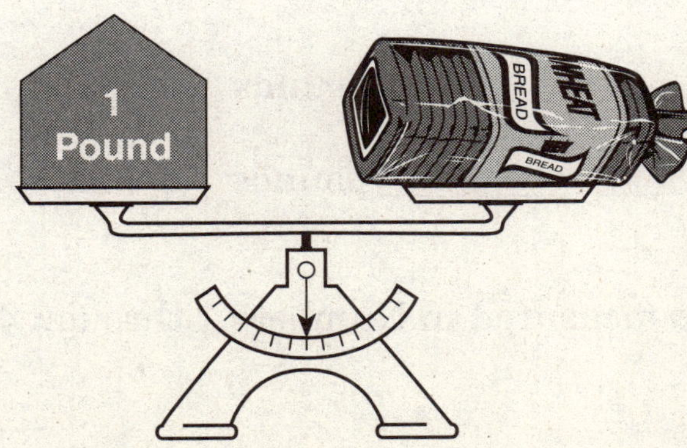

1 pound = 16 ounces
A real-world loaf of bread weighs about 1 pound.

Practice

Directions: For Numbers 1 and 2, first look at the object to estimate the weight in ounces. Then measure the weight in ounces on a scale or balance.

1. a box of crayons

   Estimate: about _______________ ounces

   Measurement: ________________ ounces

2. a stapler

   Estimate: about _______________ ounces

   Measurement: ________________ ounces

Directions: For Numbers 3 and 4, first look at the object to estimate the weight in pounds. Then measure the weight in pounds on a scale or balance.

3. a classmate

   Estimate: about _______________ pounds

   Measurement: ________________ pounds

4. a dictionary

   Estimate: about _______________ pounds

   Measurement: ________________ pounds

5. Which of the items measured in Numbers 1 through 4 is the **heaviest**?

   ____

Objectives: M.A.1, M.A.4, M.B.2

6. Grant went to the store and bought some fruits. The table below shows how much each fruit weighed.

Fruit	Weight
Papaya	5 pounds
Orange	11 ounces
Watermelon	19 pounds
Mango	20 ounces

Which weighs **less**, the orange or the mango? ____________________

7. In the real world, which object's weight would you **most likely** measure in ounces? Circle the correct answer.

8. In the real world, which object's weight would you **most likely** measure in pounds? Circle the correct answer.

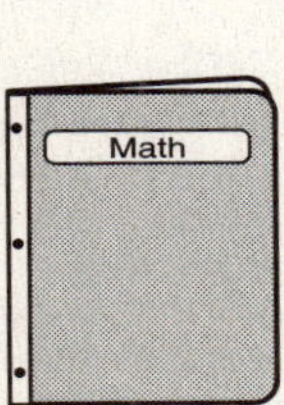

Objectives: M.A.1, M.A.3, M.B.2

Metric Units of Weight

Grams (g) and **kilograms (kg)** are two metric units used to measure weight.

Grams

Grams are used to weigh light objects.

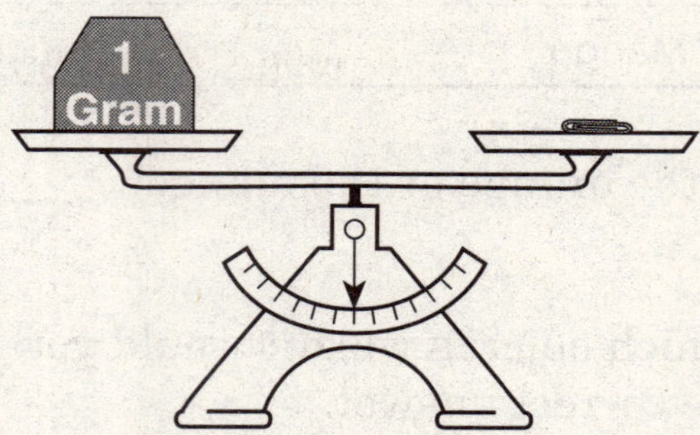

A real-world large paper clip weighs about 1 gram.

Kilograms

Kilograms are used to weigh heavier objects. A kilogram is a little more than 2 pounds.

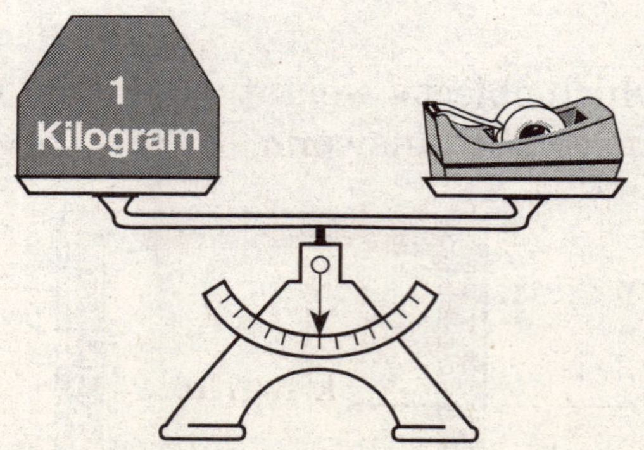

1 kilogram = 1,000 grams
A real-world tape dispenser weighs about 1 kilogram.

Objectives: M.A.1, M.A.4, M.B.2

Practice

Directions: For Numbers 1 and 2, first look at the object to guess the weight in grams. Then measure the weight in grams on a scale or balance.

1. a pencil

   Estimate: about _______________ grams

   Measurement: _______________ grams

2. a chalkboard eraser

   Estimate: about _______________ grams

   Measurement: _______________ grams

Directions: For Numbers 3 and 4, first look at the object to guess the weight in kilograms. Then measure the weight in kilograms on a scale or balance.

3. a telephone book

   Estimate: about _______________ kilograms

   Measurement: _______________ kilograms

4. a rock the size of your fist

   Estimate: about _______________ kilograms

   Measurement: _______________ kilograms

5. Which of the items measured in Numbers 1 through 4 is the **lightest**?

   ____

6. Niki visited the zoo and saw some birds. The table below shows how much each bird weighed.

Bird	Weight
Dusky Lory	155 grams
Palm Cockatoo	1 kilogram
Cockatiel	90 grams
Eclectus Parrot	450 grams

Which weighs **more**, the Dusky Lory or the Cockatiel?

____

7. In the real world, which object's weight would you **most likely** weigh in kilograms? Circle the correct answer.

8. In the real world, which object's weight would you **most likely** measure in grams? Circle the correct answer.

Mathematics Practice

1. Which weight is the **lightest**?

   Ⓐ 400 grams

   Ⓑ 40 grams

   Ⓒ 400 kilograms

   Ⓓ 40 kilograms

2. How many ounces are there in 1 pound?

   Ⓐ 4

   Ⓑ 8

   Ⓒ 16

   Ⓓ 64

3. Which is the **best estimate** for the weight of a bulldog?

   Ⓐ 6 ounces

   Ⓑ 40 pounds

   Ⓒ 40 ounces

   Ⓓ 200 pounds

4. Which object would you **most likely** weigh in kilograms?

   Ⓐ a sock

   Ⓑ a feather

   Ⓒ a grain of sand

   Ⓓ a log

5. Which of the following real-world objects weighs **about** 20 pounds?

Ⓐ 
rhinoceros

Ⓒ 
watermelon

Ⓑ 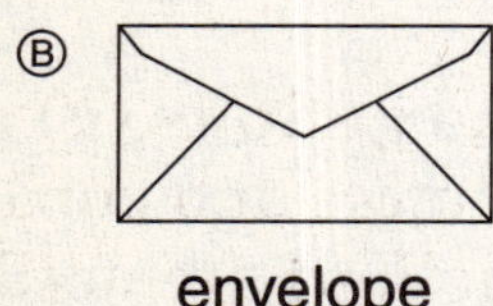
envelope

Ⓓ 
soccer ball

6. Jerry took off his shoe and put it on the scale.

How much does Jerry's shoe weigh?

Ⓐ 4 ounces
Ⓑ 6 ounces
Ⓒ 8 ounces
Ⓓ 10 ounces

Objectives: M.A.1, M.A.3, M.B.2

Lesson 15: Capacity

In this lesson, you will learn about capacity. **Capacity** is the amount something holds. You can measure capacity using U.S. customary or metric units. Round to the nearest unit in this lesson.

U.S. Customary Units of Capacity

Fluid ounces (fl oz), **cups (c)**, **pints (pt)**, **quarts (qt)**, and **gallons (g)** are five U.S. customary units used to measure capacity. Fluid ounces and cups are used to measure smaller capacities. Pints, quarts, and gallons are used to measure larger capacities.

1 cup = 8 fluid ounces

1 pint = 2 cups

1 quart = 2 pints = 4 cups

1 gallon = 4 quarts

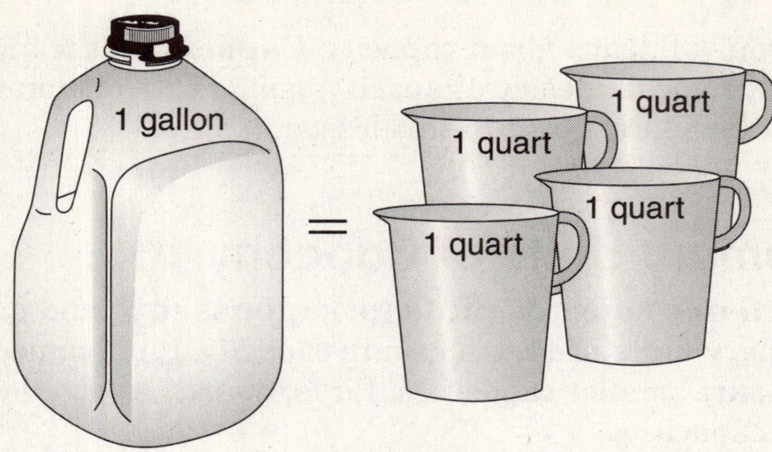

Practice

Directions: For Numbers 1 through 4, fill in the table with real-world containers, such as cups, glasses, vases, bowls, or buckets. First estimate the capacities of the containers using U.S. customary units. Then measure their capacities using U.S. customary units. Write in the units you used for your measurements.

	Container	Estimate	Measurement
1.			
2.			
3.			
4.			

Objectives: M.A.1, M.A.4, M.B.2

5. Which of the items measured in Numbers 1 through 4 has the **largest** capacity?

 ____

6. Which of the items measured in Numbers 1 through 4 has the **smallest** capacity?

 ____

7. In the real world, which object's capacity would you **most likely** measure in gallons? Circle the correct answer.

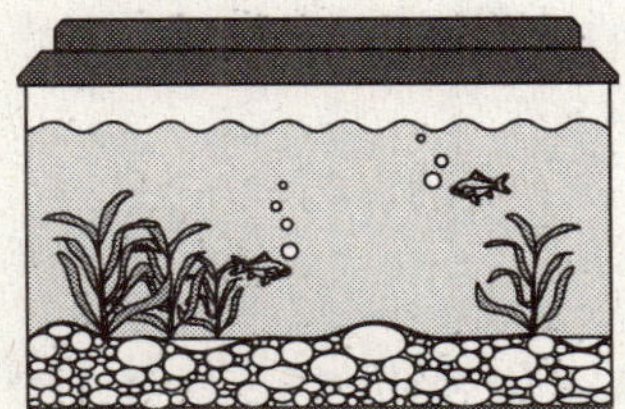

8. In the real world, which object's capacity would you **most likely** measure in fluid ounces? Circle the correct answer.

9. Which is the **best estimate** for the capacity of a bathroom sink?

 A. 5 cups

 B. 6 fluid ounces

 C. 10 pints

 D. 12 quarts

Metric Units of Capacity

Milliliters (mL) and **liters (L)** are two metric units used to measure capacity.

Milliliters

Milliliters are used to measure small amounts of liquid.

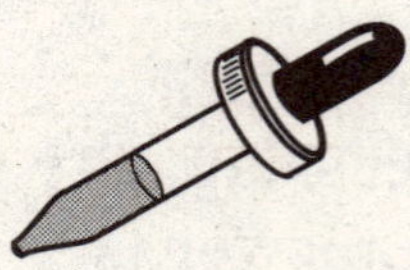

A real-world medicine dropper holds about 1 milliliter.

Liters

Liters are used to measure larger amounts of liquid. A liter is a little more than 1 quart.

1 liter = 1,000 milliliters
A real-world juice carton like this holds about 1 liter.

Practice

1. Which real-world object could hold 1 liter? Circle the correct answer.

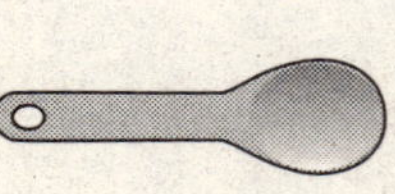

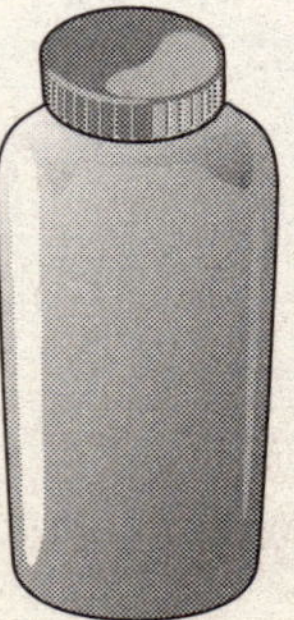

Objectives: M.A.1, M.A.4, M.B.2

Directions: For Numbers 2 through 5, fill in the table with real-world containers. First estimate the capacities of the containers using metric units. Then measure their capacities using metric units. Write in the units you used for your measurements.

	Container	Estimate	Measurement
2.			
3.			
4.			
5.			

6. Which of the items measured in Numbers 2 through 5 has the **largest** capacity?

   ____

7. Which of the items measured in Numbers 2 through 5 has the **smallest** capacity?

   ____

8. In the real world, which object's capacity would you **most likely** measure in liters? Circle the correct answer.

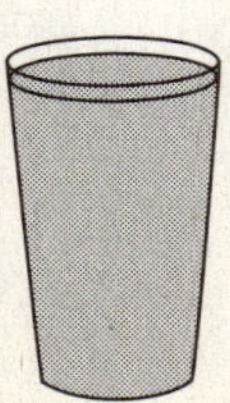

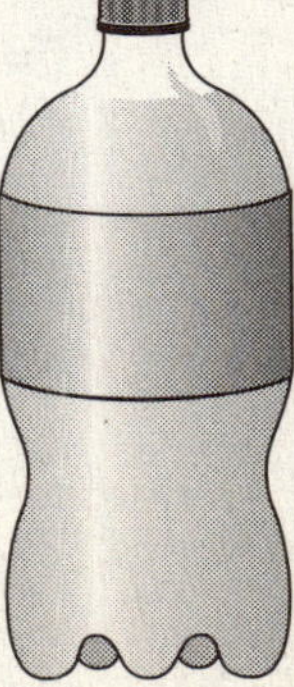

Mathematics Practice

1. Which capacity is the **largest**?

   Ⓐ 8 pints
   Ⓑ 8 quarts
   Ⓒ 8 cups
   Ⓓ 8 gallons

2. How many milliliters are there in 1 liter?

   Ⓐ 10
   Ⓑ 100
   Ⓒ 1,000
   Ⓓ 10,000

3. Sadie measured milk in the measuring cup below.

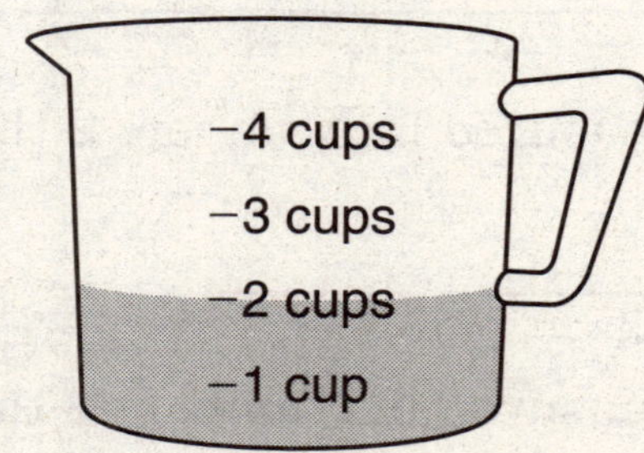

   How much milk did Sadie measure?

   Ⓐ 1 cup
   Ⓑ 2 cups
   Ⓒ 3 cups
   Ⓓ 4 cups

4. Which of the following real-world objects holds **about** 1 liter?

Ⓐ 

vase

Ⓑ 

watering can

Ⓒ 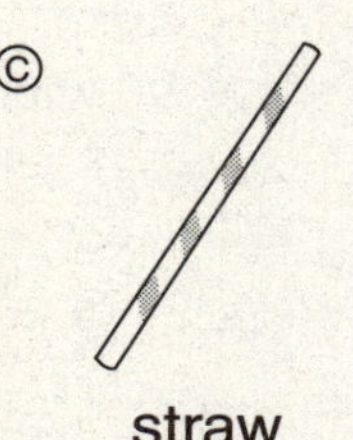

straw

Ⓓ 

fish tank

5. How many quarts are there in 1 gallon?

Ⓐ 8

Ⓑ 6

Ⓒ 4

Ⓓ 2

Objectives: M.A.1, M.A.3, M.B.3, M.B.4

Lesson 16: Time

In this lesson, you will learn about time. **Time** can be measured in **hours** and **minutes**. Clocks are used to tell time. Each day has 24 hours. There are 12 hours before noon and 12 hours after noon. To tell time **from midnight up to noon**, use A.M. To tell time **from noon up to midnight**, use P.M.

Digital Clocks

On a **digital clock**, the number before the colon (:) shows the hour. The number after the colon shows the minutes. A digital clock shows A.M. and P.M. with a little light by the A.M. or P.M.

Example

What time does the following clock show?

The light is on by the A.M. The time is 10:30 A.M.

Example

What time does the following clock show?

The light is on by the P.M. The time is 9:05 P.M.

Objectives: M.A.1, M.A.3, M.B.3, M.B.4

Analog Clocks

The numbers on an **analog** or **dial clock** show the hours. The little marks around the outside of the clock show the minutes. From one number to the next there are 5 minutes. There are 60 minutes in each hour. Most analog clocks do not show A.M. or P.M.

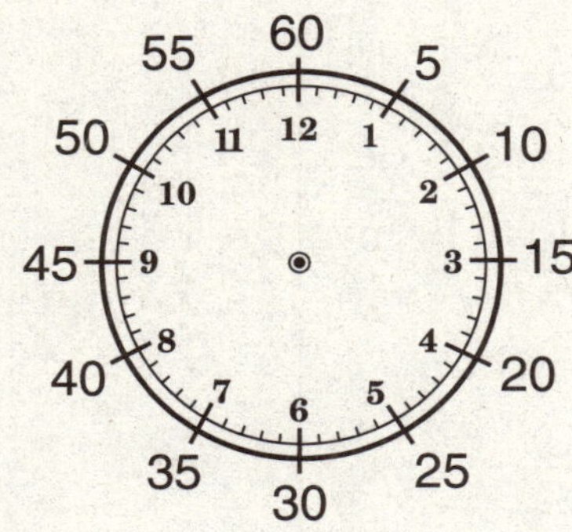

Example

What time does the following clock show? (It is between noon and midnight.)

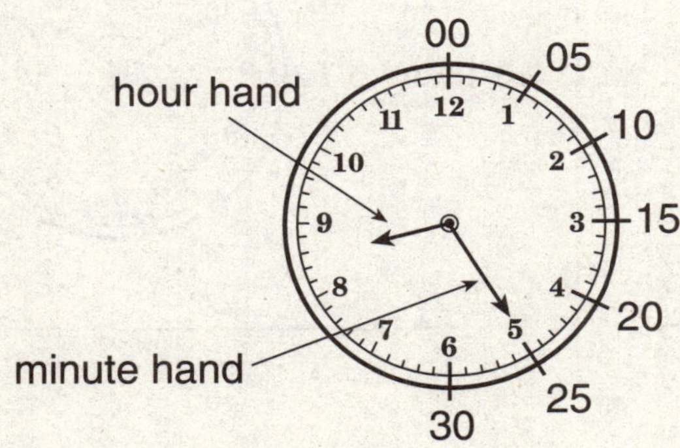

The **short hand** is called the **hour hand**. It points to the **hour**. If the hour hand is pointing between two numbers, the hour is the last number it passed as it went around the clock. In this example, the hour hand has passed 8 and is pointing between the 8 and 9. This means the hour is 8.

The **long hand** is called the **minute hand**. It points to the **minute**. It is pointing to the 5. This means it is 25 minutes past the hour.

Because it is between noon and midnight, it is P.M. time.

The time is 8:25 P.M.

Objectives: M.A.1, M.A.3, M.B.3, M.B.4

Practice

Directions: For Numbers 1 through 6, write the time shown on each clock. Be sure to include A.M. or P.M.

1. It is between noon and midnight.

____

2. It is between midnight and noon.

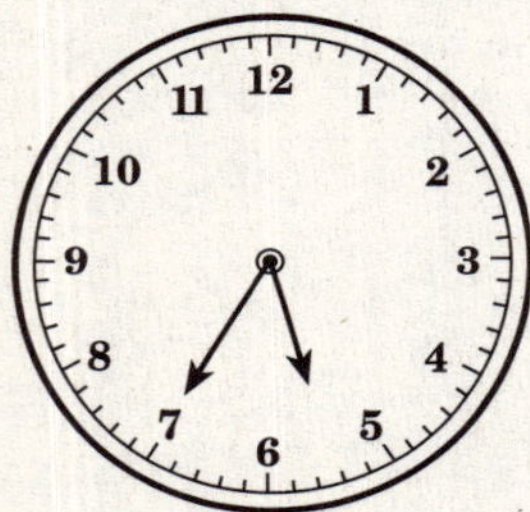

____

3.

____

4. It is between midnight and noon.

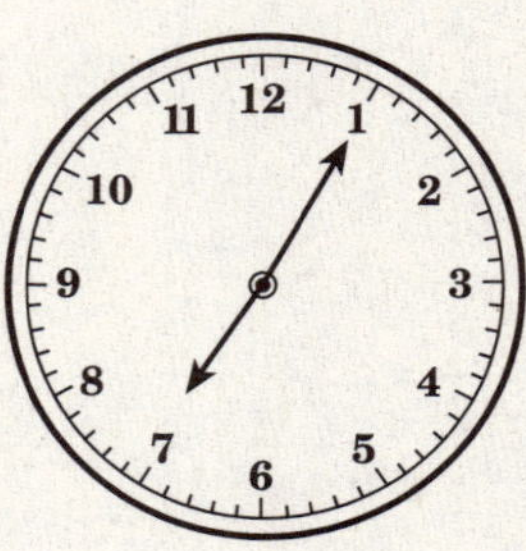

____

5. It is between noon and midnight.

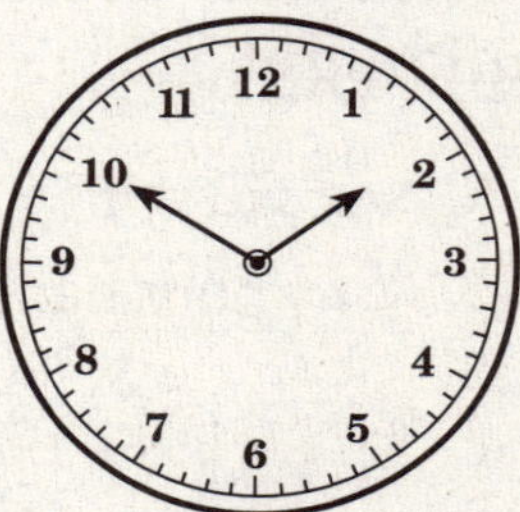

____

6.

____

Objectives: M.A.1, M.A.3, M.B.3, M.B.4

Hours, Half Hours, Quarter Hours

The following table lists some reminders that may be helpful when telling time using minutes, quarter hours, half hours, or hours.

Reminders
$\frac{1}{4}$ hour = 15 minutes
$\frac{1}{2}$ hour = 30 minutes
$\frac{3}{4}$ hour = 45 minutes
1 hour = 60 minutes

Examples

6:00 is read as:

"six o'clock"

6:15 can be read as:

"six fifteen"

"fifteen minutes after six"

"a quarter after six"

6:30 can be read as:

"six thirty"

"thirty minutes after six"

"half past six"

6:45 can be read as:

"six forty-five"

"forty-five minutes after six"

"fifteen minutes before **seven**"

"a quarter to **seven**"

Practice

Directions: For Numbers 1 through 6, write the time shown on the clock in two different ways. Be sure to include A.M. or P.M.

1. It is between noon and midnight.

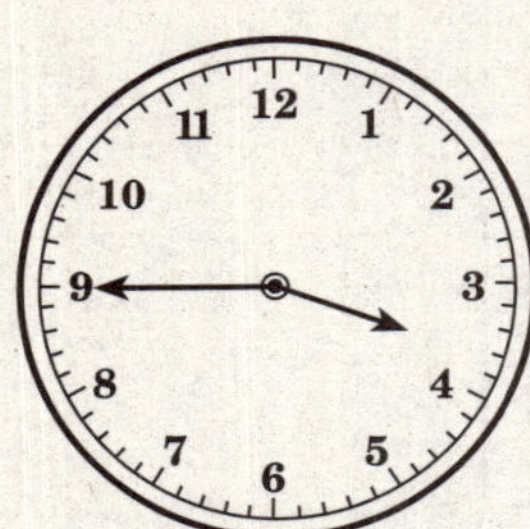

____

____

2. It is between midnight and noon.

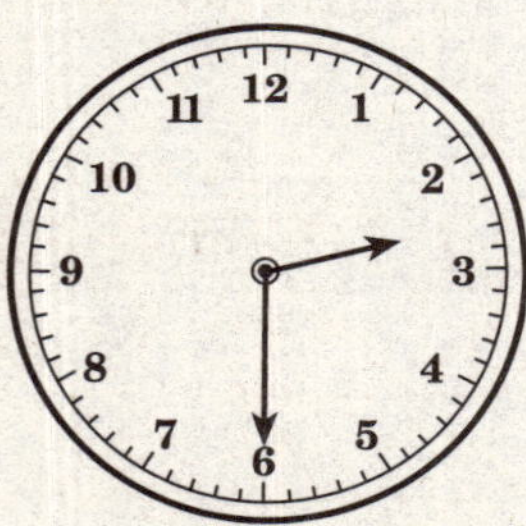

____

____

3. 

____

____

4. It is between noon and midnight.

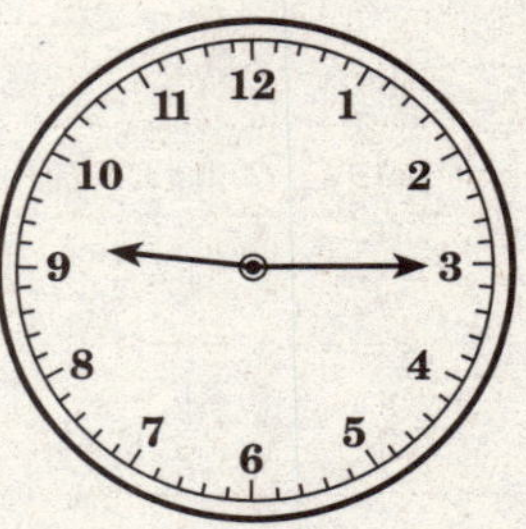

____

____

5. It is between midnight and noon.

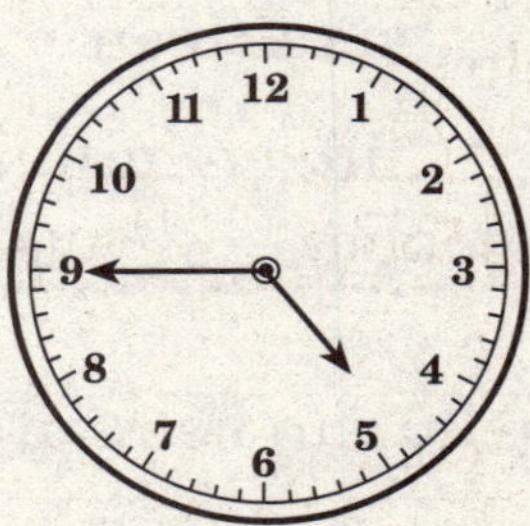

____

____

6. 

____

____

Objectives: M.A.1, M.A.3, M.B.3, M.B.5

Days, Weeks, Months, and Years

Calendars are used to tell time in **days**, **weeks**, **months**, and **years**.

There are **7 days** in **1 week**. The days of the week are shown in the calendar below.

JULY						
Sunday	Monday	Tuesday	Wednesday	Thursday	Friday	Saturday
		1	2	3	4	5
6	7	8	9	10	11	12
13	14	15	16	17	18	19
20	21	22	23	24	25	26
27	28	29	30	31		

There are **12 months** in **1 year**. There are **28** to **31 days** in **1 month**. The months of the year and the number of days in each month are shown in the tables below.

Month	Number of Days
January	31
February	28
March	31
April	30
May	31
June	30

Month	Number of Days
July	31
August	31
September	30
October	31
November	30
December	31

There are **365 days**, **52 weeks**, or **12 months** in **1 year**.

TIP: Sometimes February has 29 days. This happens every 4 years (leap year). The year 2004 was a leap year.

Objectives: M.A.1, M.A.3, M.B.3, M.B.5

Practice

1. Complete the calendar to show this month.

Month: ______						
Sunday	Monday	Tuesday	Wednesday	Thursday	Friday	Saturday

2. How many days are in this month? ______

3. What day of the week is today? ______

4. What is the date of the last Friday of this month?

______

5. What is the date of the second Tuesday of this month?

______

6. On what day is the 4th of this month? ______

7. On what day is the 21st of this month? ______

Objectives: M.A.1, M.A.3, M.B.3, M.B.5

8. Darian's brother is 1 year old. How many **months** old is Darian's brother?

______

9. Nick has been at camp for 2 weeks. How many **days** has Nick been at camp?

______

Directions: Use the calendars below to answer Numbers 10 through 12.

FEBRUARY						
Sun.	Mon.	Tues.	Wed.	Thurs.	Fri.	Sat.
				1	2	3
4	5	6	7	8	9	10
11	12	13	14	15	16	17
18	19	20	21	22	23	24
25	26	27	28			

MARCH						
Sun.	Mon.	Tues.	Wed.	Thurs.	Fri.	Sat.
				1	2	3
4	5	6	7	8	9	10
11	12	13	14	15	16	17
18	19	20	21	22	23	24
25	26	27	28	29	30	31

10. Valentine's Day is on February 14. On what day of the week is Valentine's Day?

______

11. Jane's birthday is on the last day of February. What is the date of Jane's birthday?

A. February 28

B. February 29

C. February 30

D. February 31

12. Anna went to the art museum on the third Monday in March. On what date did Anna go to the art museum?

A. March 11

B. March 12

C. March 18

D. March 19

Mathematics Practice

1. Which clock shows 11:25?

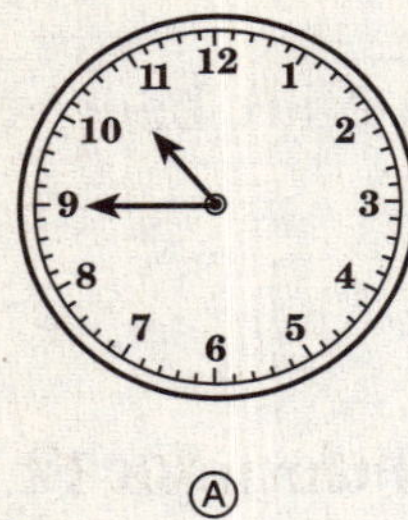
Ⓐ

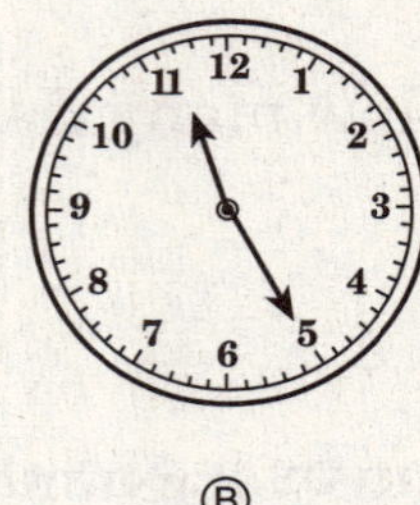
Ⓑ

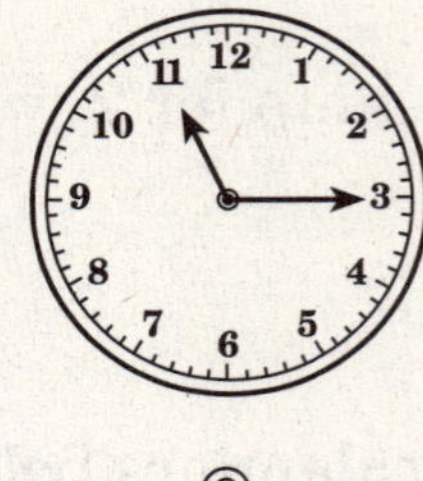
Ⓒ

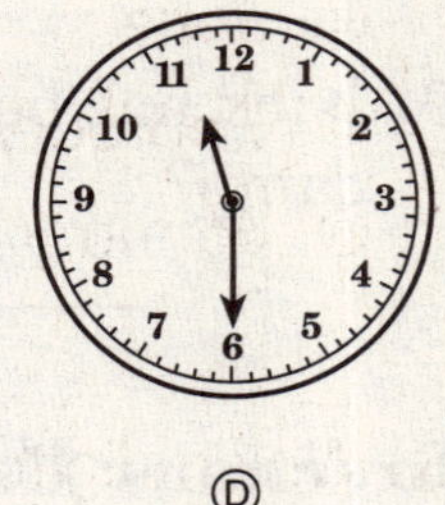
Ⓓ

2. How many months are in 1 year?

Ⓐ 4

Ⓑ 10

Ⓒ 12

Ⓓ 365

3. Which is a way of writing the time shown on the clock below?

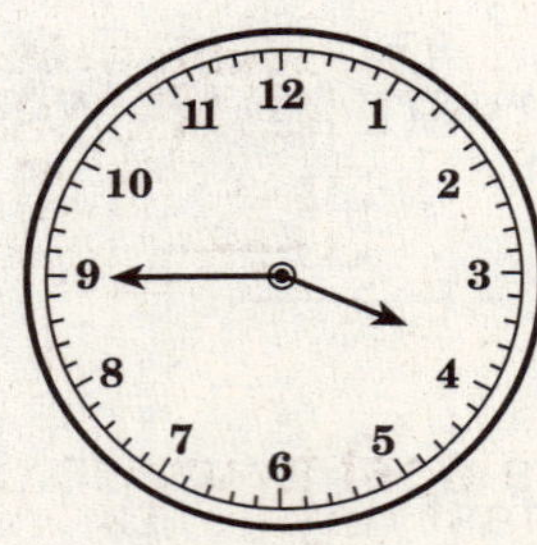

Ⓐ a quarter to four

Ⓑ a quarter after three

Ⓒ a quarter after four

Ⓓ half past three

4. Beverly's birthday month has 31 days. Which of these months could be Beverly's birthday month?

   Ⓐ November

   Ⓑ June

   Ⓒ February

   Ⓓ January

5. Which digital clock shows the same time as the following analog clock?

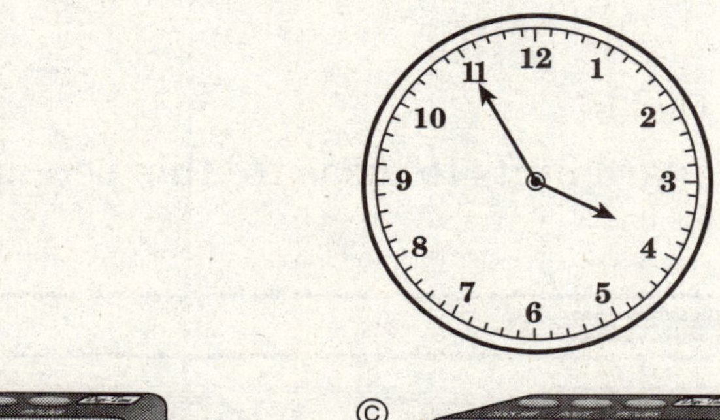

   Ⓐ AM PM 3:25

   Ⓑ AM PM 3:55

   Ⓒ AM PM 4:05

   Ⓓ AM PM 4:25

6. Which unit of time would be **best** to use to describe how long you are at school each day?

   Ⓐ seconds

   Ⓑ minutes

   Ⓒ hours

   Ⓓ weeks

Lesson 17: Measurement Concepts

In this lesson, you will learn about different ways to measure length, weight, and capacity. You will also review measurement tools.

Nonstandard Units

U.S. customary and metric units are also called **standard units**. They mean the same thing to everyone, everywhere. But it is also possible to measure with **nonstandard units**.

Example

Instead of using standard units to measure this pen, use paper clips.

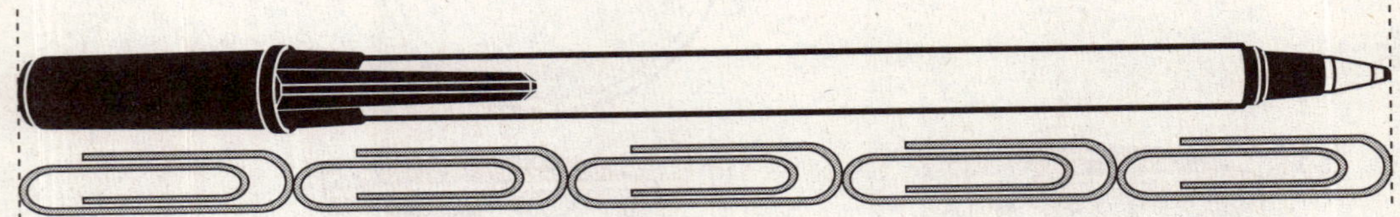

The pen is about 5 paper clips long.

Example

Instead of using standard units to weigh this roll of film, use checkers.

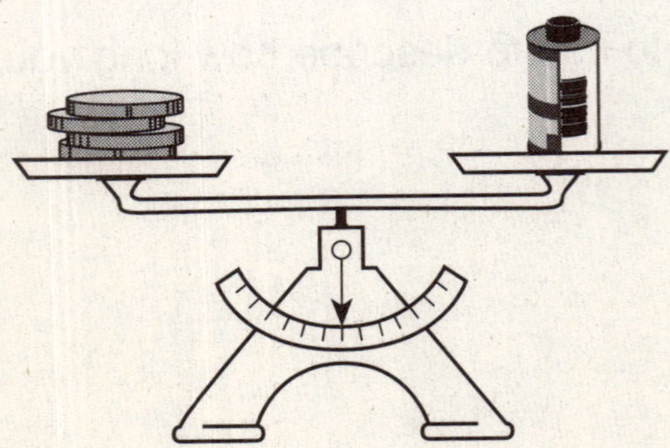

The roll of film weighs 4 checkers.

Objectives: M.A.2, M.B.1

Example

Instead of using standard units to find the capacity of a water pitcher, use drinking glasses.

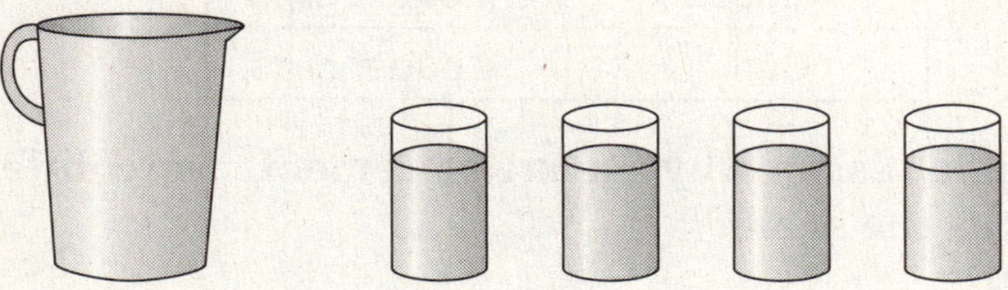

This water pitcher holds 4 drinking glasses of water.

Practice

1. Look around your classroom. Find 3 objects. Measure the length of each object using the same nonstandard unit. What nonstandard unit will you use to measure?

   ____

   Fill in the following table with your measurements.

Object	Length

   Write the objects in order from **longest** to **shortest**.

   ____

2. Two students measured the length of the pen on page 188. The following table shows their measurements.

Name	Measurement
Janice	17 paper clips
Carl	4 paper clips

What is one possible reason why the measurements were different even though the pen was the same?

____

3. Tara, Gloria, and Freddy weighed different objects in their classroom using the same checkers. The following table shows their measurements.

Object	Weight
Pencil	2 checkers
Calculator	9 checkers
Tape	7 checkers

Write the objects in order from **heaviest** to **lightest**.

____

4. Pete, Mark, and Tiffany each emptied their juice boxes into glasses that were the same size. The following table shows their measurements.

Person	Capacity
Pete	2 glasses
Mark	3 glasses
Tiffany	1 glass

Write the children in order by how much their juice boxes held from **least** to **greatest**.

____

Objectives: M.B.3

Measurement Tools

You use tools to measure in U.S customary and metric units.

A ruler is used to measure shorter lengths. A tape measure is used to measure longer lengths.

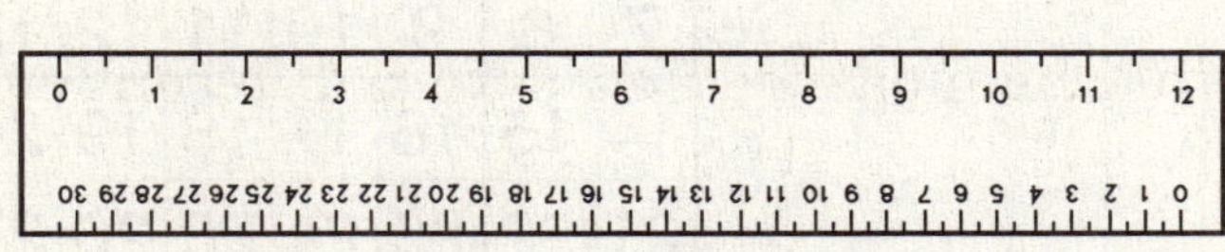

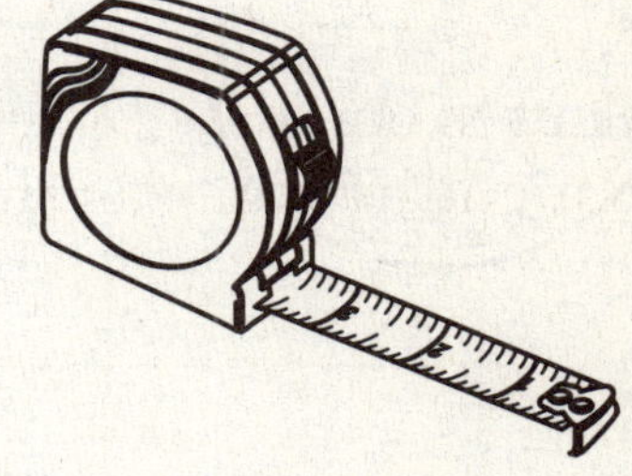

Scales and balances are used to measure weight. Three types of scales are shown below.

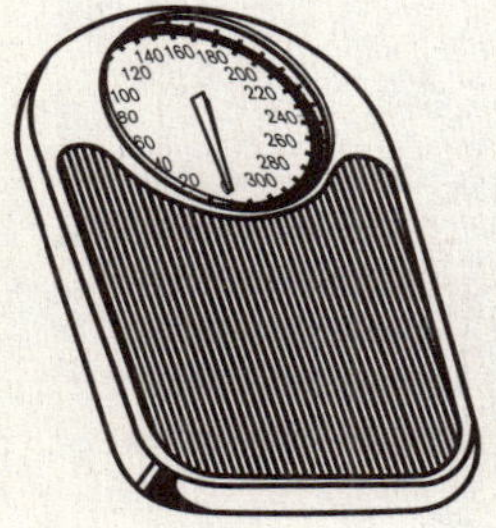

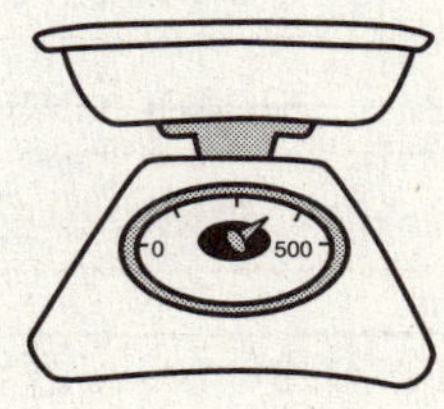

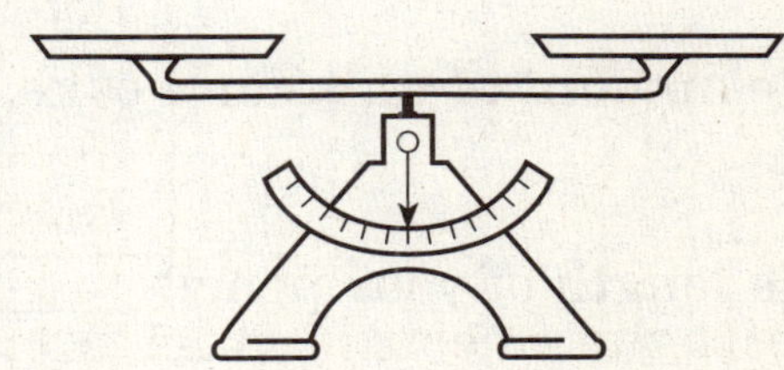

Capacity is usually measured using a measuring cup. Measuring cups come in many different sizes.

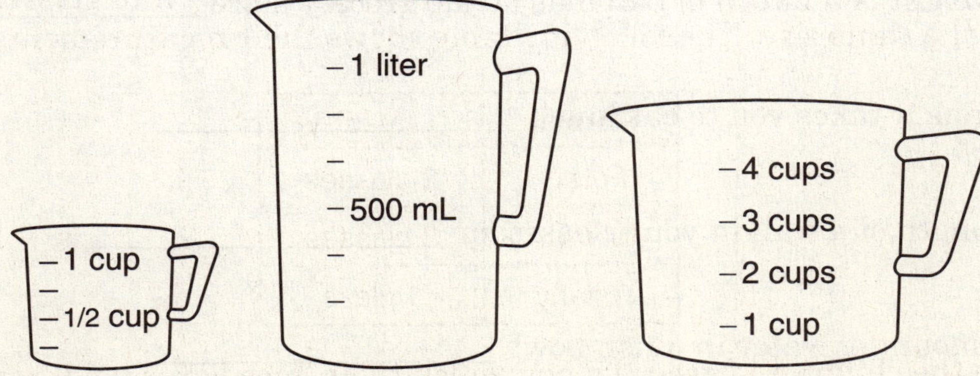

Time is measured using clocks and calendars.

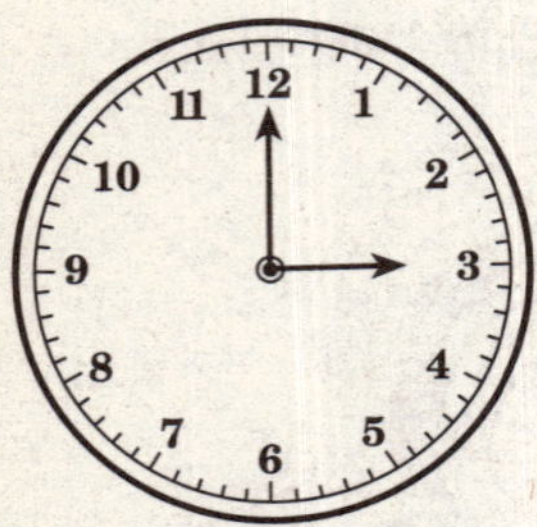

May

Sun.	Mon.	Tues.	Wed.	Thurs.	Fri.	Sat.
	1	2	3	4	5	6
7	8	9	10	11	12	13
14	15	16	17	18	19	20
21	22	23	24	25	26	27
28	29	30	31			

Practice

Directions: For Numbers 1 through 8, write which tool would be best to use to measure each item.

1. the amount of milk for a cake ____________________

2. the length of your pencil ____________________

3. the length of a vacation ____________________

4. the weight of a bunch of bananas at the grocery store ____________________

5. the time it takes you to eat lunch ____________________

6. the length of a wall in your classroom ____________________

7. the amount of water in a fish bowl ____________________

8. the amount of weight you have gained this year ____________________

Mathematics Practice

1. Molly wants to buy new curtains for her windows. Which tool would be **best** to use to make sure the curtains she buys are long enough?

Ⓐ

Ⓒ
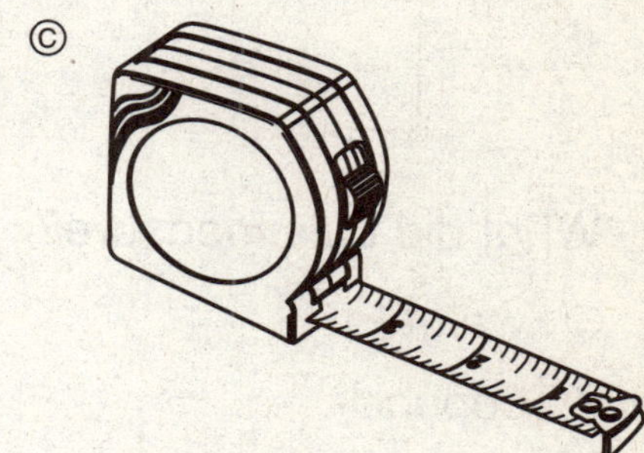

Ⓑ
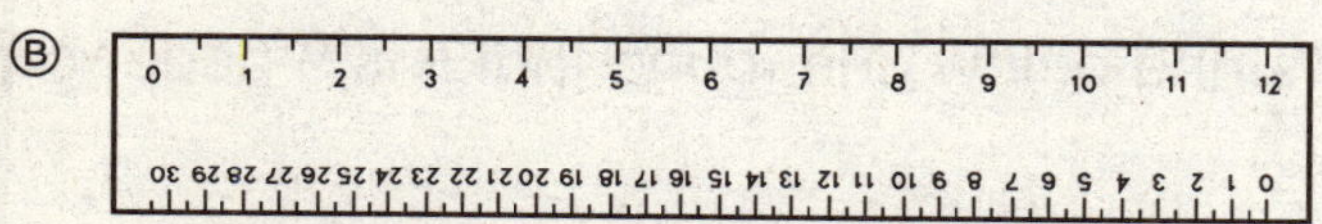

Ⓓ
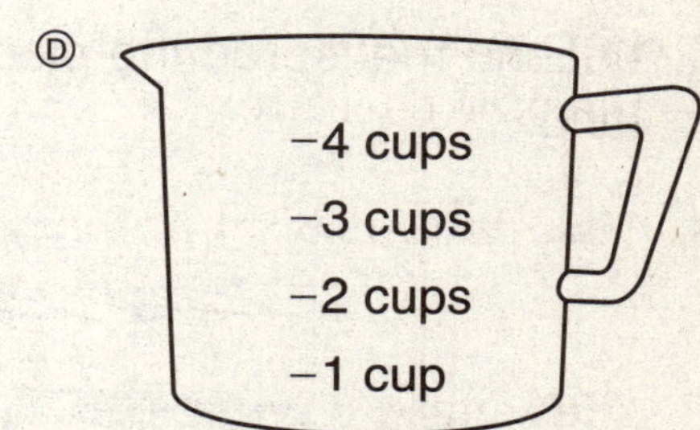

2. Which of the following would be the **best** nonstandard unit to measure the amount of water in a bucket?

Ⓐ paper clips

Ⓑ crayons

Ⓒ checkers

Ⓓ glasses

3. Luis used the following tool to measure his math book.

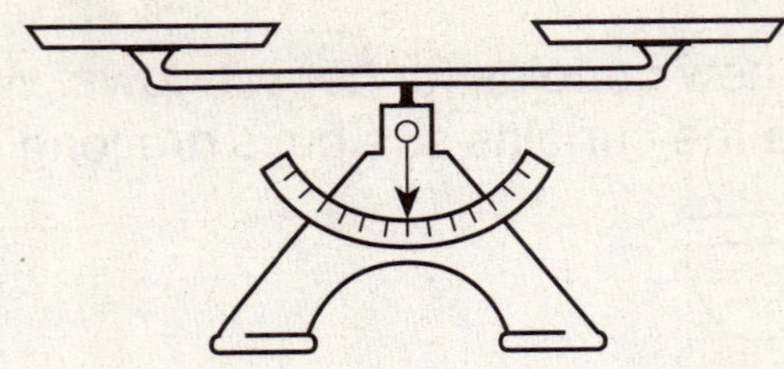

What did Luis measure?

Ⓐ weight

Ⓑ capacity

Ⓒ length

Ⓓ time

4. Latisha measured the piece of string below and found that it is 4 paper clips long.

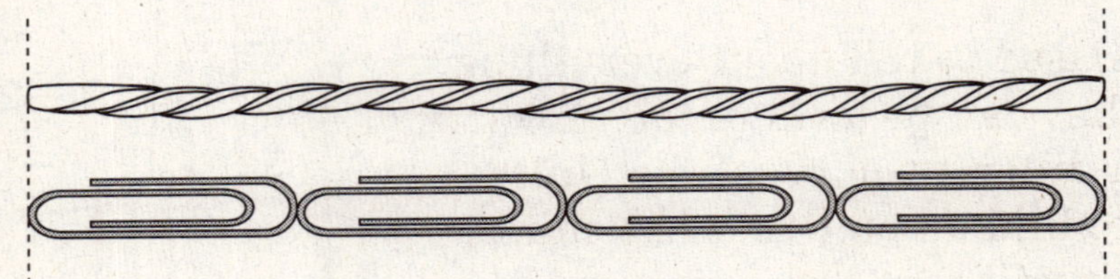

Desmond measured the same piece of string using a different nonstandard unit. It took about 9 of Desmond's units to measure the string. Which nonstandard unit did Desmond use?

Ⓐ 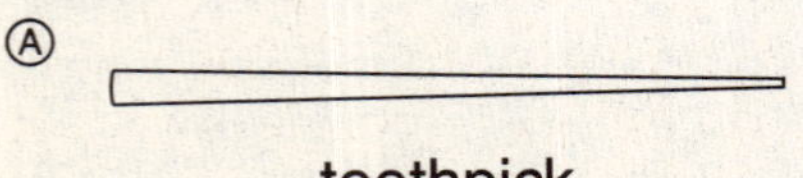

toothpick

Ⓑ 

dime

Ⓒ nail

Ⓓ 

eraser

Unit 5

Data Analysis and Probability

Numbers are used to tell about everything around us. They can be used to tell how many people live in different parts of the United States, how to find a hotel, or what chance there is that it will rain tomorrow.

In this unit, you will read charts, pictographs, and bar graphs. You will also get to make some graphs yourself. Finally, you will do probability experiments and tell how likely something is to happen.

In This Unit

Lesson 18: Data Analysis

In this lesson, you will learn about ways to collect and show information, or **data**.

Collecting Data

You can collect data by asking questions or conducting a survey. You may get different information depending on who you ask or what questions you ask. You also need to be sure that the questions you ask tell you the information you want to know.

Example

Austin wants to find out what second graders do after school. He wrote the following survey questions. Which of the questions will **not** help Austin find out what second graders do after school?

1. What is your favorite thing to do after school?
2. What chores do you have to do at home after school?
3. Do you read books at home after school?
4. What time do you get home from school?

The last question will not help Austin find the information he wants because it asks about time instead of asking about what the students do after school.

You can also collect data by observing people or things around you.

Example

Austin wants to collect data about the color of his friends' hair. He wrote down the colors he saw.

brown, brown, red, blonde, black, blonde, black, brown, brown

Austin observed that 4 of his friends have brown hair, 1 has red hair, 2 have blonde hair, and 2 have black hair.

Objectives: D.A.1

Practice

1. Think of a survey question you could use to find something out about your classmates and their lunches.

____

____

____

2. Now ask 10 of your classmates your survey question. Write down your results on the following lines.

____

____

____

3. Think of data you could collect by observing people or things around you.

____

____

____

4. Now collect the data by observing 10 people or things. Write down your results on the following lines.

____

____

____

Charts

Charts often use tally marks to show how many of something there are. The **key** shows you what the tally marks stand for.

Example

Janice asked each of her classmates what his or her favorite fruit is. Each time one of her classmates answered, she put a tally mark by that type of fruit. At the end, she counted the tally marks for each kind of fruit. The following chart shows her results.

Favorite Fruit

Fruit	Tally	Count
Banana	𝍸 |	6
Strawberry	𝍸	5
Cherry	|	1
Apple	||	2
Orange	|	1

KEY
| = 1 student vote
𝍸 = 5 student votes

How many of Janice's classmates chose strawberries?

Look under the "Fruit" column for a strawberry. Look across that row. The table shows that 5 of Janice's classmates chose strawberries.

Objectives: D.A.2, D.A.3

Practice

Directions: Use the following information to answer Numbers 1 through 5.

Travis asked his friends which U.S. cities they have visited. His results are shown in the chart below.

U.S. Cities Visited

City	Tally	Count
Chicago	IIII	4
New York	III	3
Los Angeles	~~IIII~~ IIII	9
Atlanta	~~IIII~~	5

1. Which city have the **most** people visited? ____________________

2. Which city have the **least** people visited? ____________________

3. How many people have visited Chicago? ____________________

4. How many more people have visited Los Angeles than Chicago?

____

5. How many more people have visited Atlanta than New York?

____

6. Ask your classmates which of the following places they would most like to go on a vacation: the mountains, an ocean, a desert, or a lake. Tally each answer in the chart below. Then write in the total count for each place.

Places to Visit on Vacation

Place	Tally	Count
Mountains		
Ocean		
Desert		
Lake		

KEY
| = 1 student
𝍸 = 5 students

Directions: Use the chart you made in Number 6 to answer Numbers 7 through 11.

7. How many students chose the mountains? ____________

8. How many students chose a desert? ____________

9. What is the difference between the number of students that chose an ocean and the number of students that chose a lake?

____

10. Order the places from the place that the **greatest** number of people said they would like to go to the place that the **least** number of people said they would like to go.

____

11. Compare your chart with a classmate's chart. Are they the same? Explain why they are the same or why they are different.

____

____

Objectives: D.A.2, D.A.3

Pictographs

Pictographs (or picture graphs) use pictures to show how many of something there are. Most pictographs have a key that shows what each picture stands for.

Example

Janice made the following pictograph from the chart on page 198.

Favorite Fruit

Banana	
Strawberry	
Cherry	
Apple	
Orange	

KEY
= 1 student

How many of Janice's classmates chose bananas?

The pictograph shows 6 pictures in the banana row. The key shows that each picture stands for 1 student. So, 6 students chose bananas as their favorite fruit.

Practice

Directions: Use the following information to answer Numbers 1 through 4.

The pictograph below shows the number of [book] read by second graders each week.

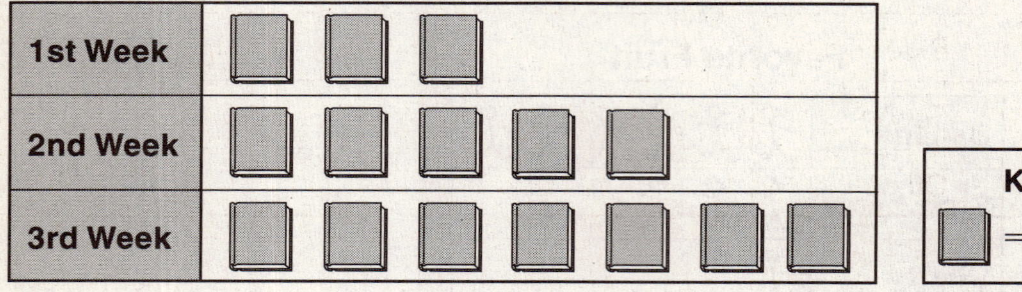

1. How many more books did the second graders read in the 3rd week than in the 1st week?

 ____

2. Add the number of books read each week. How many total books did the second graders read in the three weeks?

 ____

3. During which week did the second graders read 5 books? ____________

4. Which statement about the information in the pictograph is **not true**?

 A. Three books were read in the 1st week.

 B. The greatest number of books was read in the 3rd week.

 C. Two more books were read in the 1st week than were read in the 2nd week.

 D. Two more books were read in the 3rd week than were read in the 2nd week.

5. Look at the eye colors of 20 people. Mark each person's eye color in the following chart with a tally. Then write in the total count for each eye color.

Eye Colors

Color of Eyes	Tally	Count
Brown		
Blue		
Green		
Other		

KEY
| = 1
卌 = 5

Directions: Use the chart you made in Number 5 to answer Numbers 6 through 9.

6. Make a pictograph of the information.

Eye Colors

Brown	
Blue	
Green	
Other	

KEY
(half eye) = 1 person
(eye) = 2 people

7. Which eye color did the **most** people have? ________________

8. Which eye color did the **fewest** people have? ________________

9. If you look at the eye color of one more person, what eye color do you think he or she will have? Explain your answer.

____

____

Bar Graphs

Bar graphs use bars to compare information. The bars can go up from the bottom to the top or across from the left to the right. By looking at the heights or lengths of the bars, you can quickly compare the information in a bar graph.

Example

Janice made the following bar graph from the information in the chart on page 198.

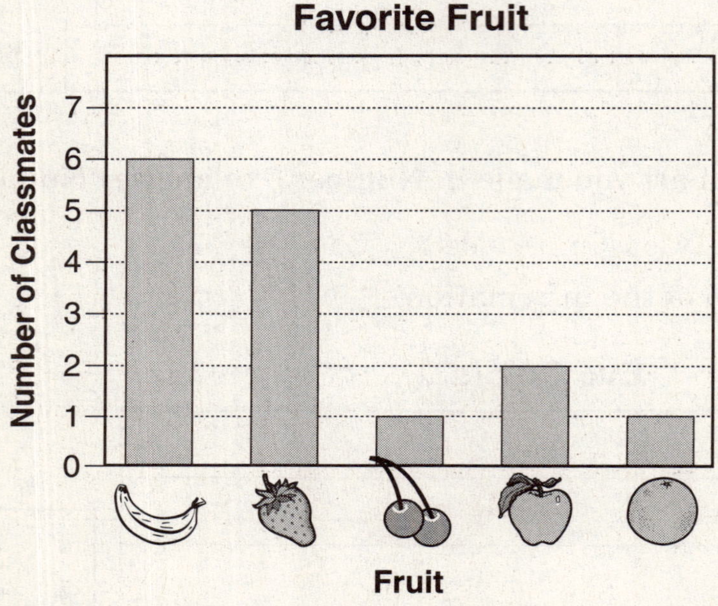

How many of Janice's classmates chose apples?

Look along the bottom of the graph until you find the picture of the apple. Put your finger on the top of the bar above the apple. Trace the line to the left to see how many students chose apples. Your finger should land on the 2. So, 2 of Janice's classmates chose apples as their favorite fruit.

Objectives: D.A.2, D.A.3

Practice

Directions: Use the following information to answer Numbers 1 through 5.

Maya visited an amusement park. The bar graph below shows how many times she rode each ride.

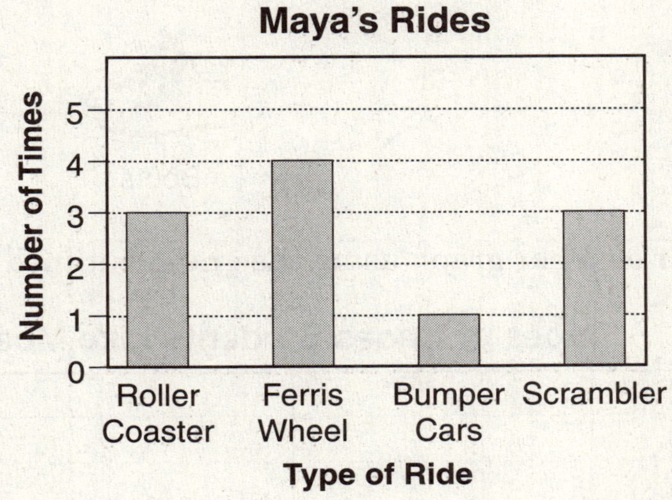

1. Which ride did Maya go on the **most**? ______________________________

2. Which ride did Maya go on the **least**? ______________________________

3. How many times did Maya ride the roller coaster? ________

4. How many more times did Maya ride the scrambler than the bumper cars?

   ____

5. How many times did Maya go on rides in all? ________

6. Ask your classmates which type of shoe they like to wear the most. Then record the numbers in the following spaces.

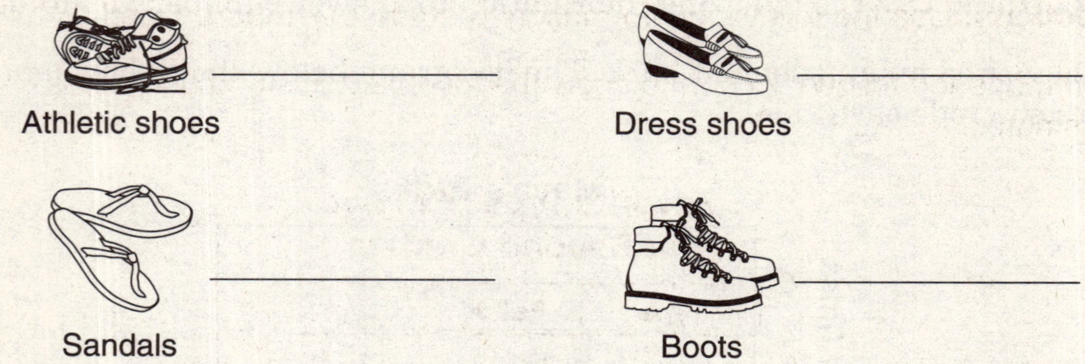

Now make your own bar graph using the graph outlined below.

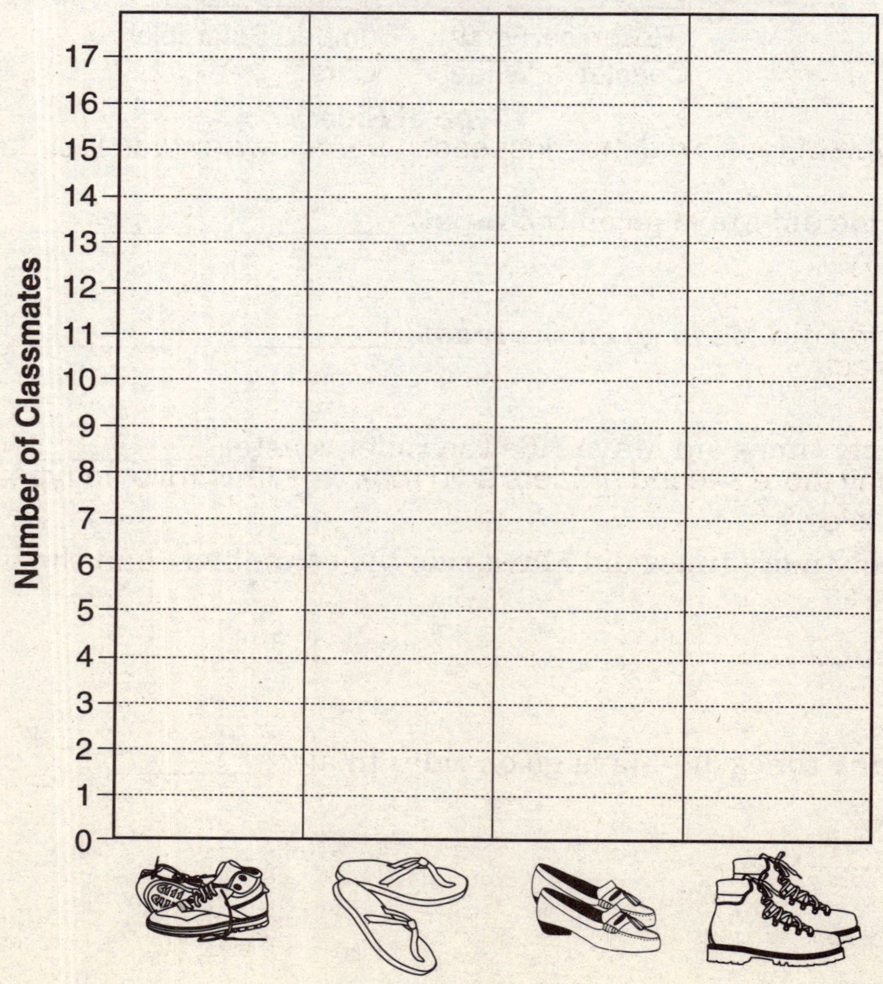

Mathematics Practice

Directions: Use the following information to answer Numbers 1 and 2.

Scott made the following chart to show the favorite types of music of his classmates.

Favorite Type of Music for Second Graders

Type	Tally	Count
Country	𝍸 III	8
Rock	𝍸	5
Blues	IIII	4
Pop	𝍸 II	7
Classical	II	2

1. Which type of music did the most second graders say was their favorite?

   Ⓐ rock

   Ⓑ country

   Ⓒ pop

   Ⓓ classical

2. How many more second graders said rock was their favorite type of music than classical?

   Ⓐ 3

   Ⓑ 4

   Ⓒ 5

   Ⓓ 7

Directions: Use the following information to answer Numbers 3 and 4.

Miss Bliss made the following pictograph to show her students' favorite colors.

Students' Favorite Colors

Red	
Yellow	
Blue	
Green	

3. How many students named blue as their favorite color?

   Ⓐ 12
   Ⓑ 8
   Ⓒ 6
   Ⓓ 4

4. How many more students named green as their favorite color than named red?

   Ⓐ 9
   Ⓑ 6
   Ⓒ 5
   Ⓓ 3

Directions: Use the following information to answer Numbers 5 and 6.

The bar graph below shows the favorite ice cream flavors of Molly's classmates.

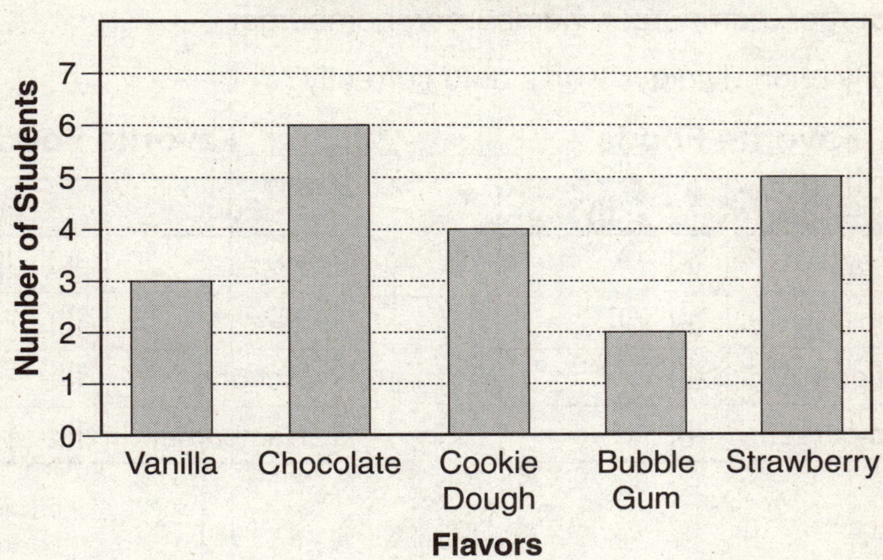

5. Which flavor was chosen by the **most** students?

   Ⓐ chocolate

   Ⓑ cookie dough

   Ⓒ bubble gum

   Ⓓ strawberry

6. How many more students chose strawberry than vanilla?

   Ⓐ 5

   Ⓑ 3

   Ⓒ 2

   Ⓓ 1

7. Karl asked his classmates what their favorite food is. The following list shows their answers.

   pizza, pizza, pizza, pizza, pizza, pizza, pizza, pizza, taco, taco, taco, taco, taco, hot dog, hot dog, hot dog, hot dog, hamburger, hamburger, hamburger, hamburger, hamburger, hamburger

   Which tally chart displays Karl's data correctly?

Ⓐ **Favorite Foods**

Food	Tally
Pizza	𝍸
Taco	𝍸 𝍷𝍷𝍷
Hot dog	𝍷𝍷𝍷𝍷
Hamburger	𝍸 𝍷

Ⓑ **Favorite Foods**

Food	Tally
Pizza	𝍸 𝍷
Taco	𝍷𝍷𝍷𝍷
Hot dog	𝍸
Hamburger	𝍸 𝍷𝍷𝍷

Ⓒ **Favorite Foods**

Food	Tally
Pizza	𝍸 𝍷𝍷𝍷
Taco	𝍷𝍷𝍷𝍷
Hot dog	𝍸
Hamburger	𝍸 𝍷

Ⓓ **Favorite Foods**

Food	Tally
Pizza	𝍸 𝍷𝍷𝍷
Taco	𝍸
Hot dog	𝍷𝍷𝍷𝍷
Hamburger	𝍸 𝍷

Objectives: D.B.1

Lesson 19: Probability

In this lesson, you will review probability. **Probability** is how likely it is that something will happen.

Likelihood of Events

You can predict if something is **most likely**, **least likely**, or **equally likely** to happen.

Example

Look at the bag of marbles below. Tyler is going to pick one marble from the bag without looking. What is the likelihood of each kind of marble being picked?

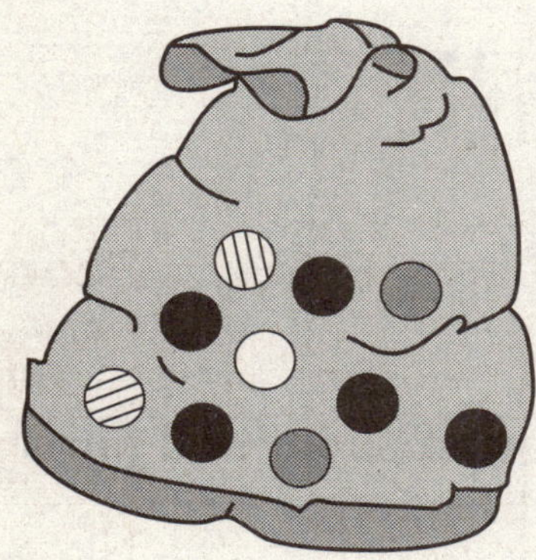

There are 5 black, 2 gray, 2 striped, and 1 white marble in the bag.

Since there are more black marbles than any other kind of marble, a black marble is **most likely** to be picked.

Since there is the same number of gray and striped marbles, they are **equally likely** to be picked.

Since there are fewer white marbles than any other kind of marble, a white marble is **least likely** to be picked.

Practice

Directions: Sasha will pick one of these balls from a box without looking. For Numbers 1 and 2, tell if the ball shown is **most likely**, **least likely**, or **equally likely** to be picked. Circle your answer.

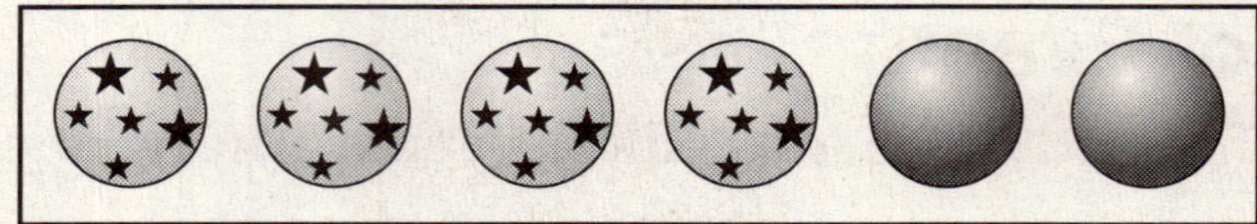

1. 

   most likely least likely equally likely

2. 

   most likely least likely equally likely

3. Aaron wrote the following numbers on slips of paper and put them in a bag. He will pick one slip of paper without looking.

   2 1 3 2 1 2 4 1 4

   Which numbers are **equally likely** to be picked? ________________

Objectives: D.B.1

4. Tell how likely it is to pick a ● out of the bag without looking. Circle your answer.

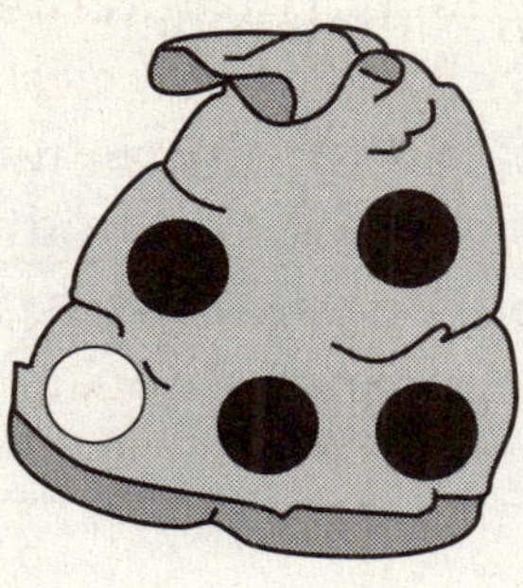

most likely   least likely   equally likely

5. On which spinner is the ➤ **least likely** to stop on the white area? Circle your answer.

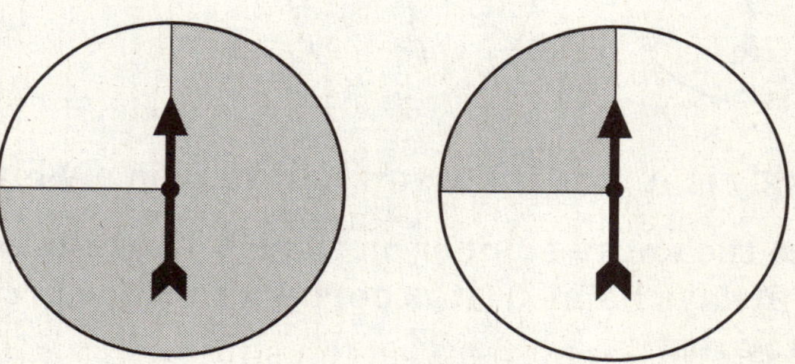

6. Which spinner shows that the ➤ is **equally likely** to stop on the white area or the gray area? Circle your answer.

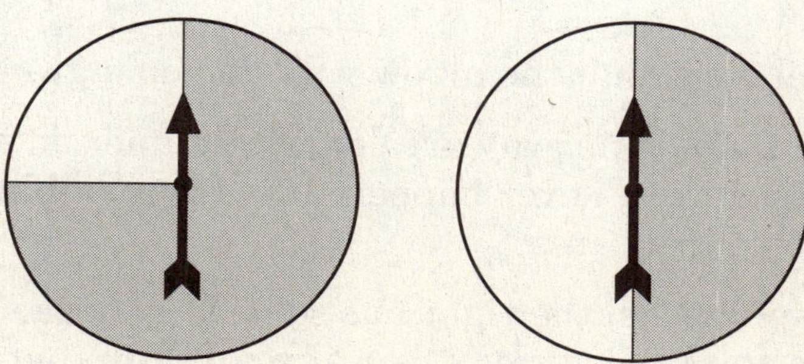

Certain, Probable, and Impossible Events

An event that is **certain** is one that will **always** happen. An event that is **impossible** is one that will **never** happen.

Some events are neither certain nor impossible. For example, if you roll a number cube, you might roll a 3, but you might not. Getting a 3 is neither certain nor impossible. **Probable** and **not probable** describe the likelihood of events that are neither certain nor impossible.

Example

Bert is going to spin each of the spinners below.

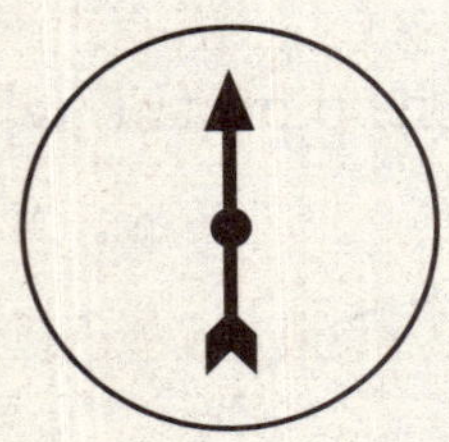

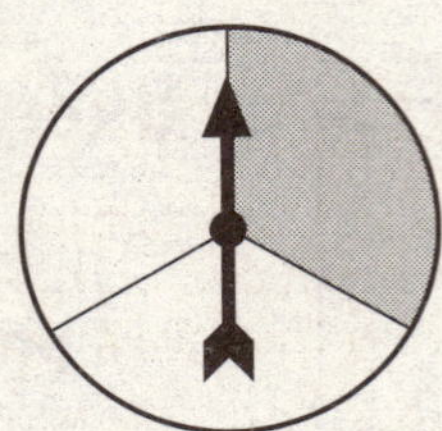

How likely is it that the arrow will stop on a white area?

Since the entire spinner on the left is white, this is an event that will always happen. It is **certain** that the arrow will stop on a white area.

The spinner on the right has both white and gray areas, so it is neither certain nor impossible. Since more than half of the spinner is white, it is **probable** that the arrow will stop on a white area.

How likely is it that the arrow will stop on a gray area?

Since there is no gray area on the spinner on the left, this is an event that can never happen. It is **impossible** for the arrow to stop on a gray area.

The spinner on the right has some gray area. Since less than half of the spinner is gray, it is **not probable** that the arrow will stop on a gray area.

Objectives: D.B.1

Practice

Directions: For Numbers 1 through 4, write *certain*, *impossible*, *probable*, or *not probable* for each event.

1. the arrow landing on green

____

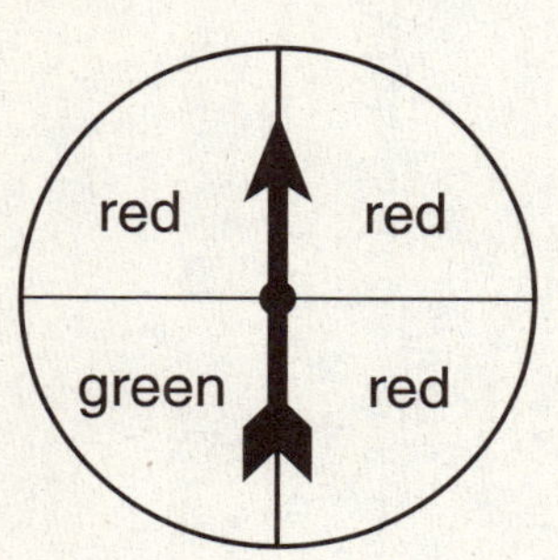

2. picking a penny out of this pocket

____

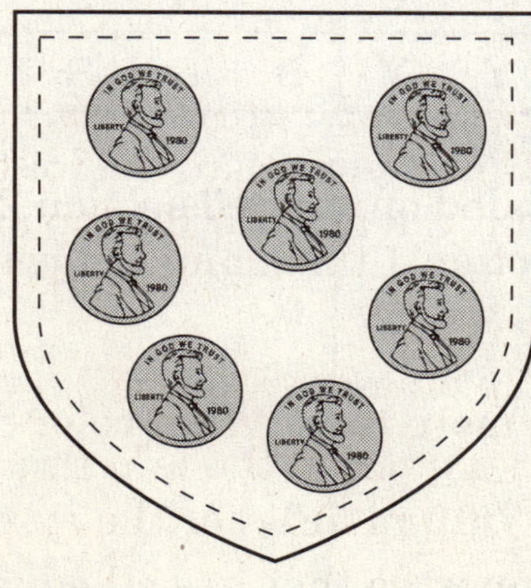

3. rolling a 7 on a number cube numbered 1 through 6

____

4. picking a striped marble

____

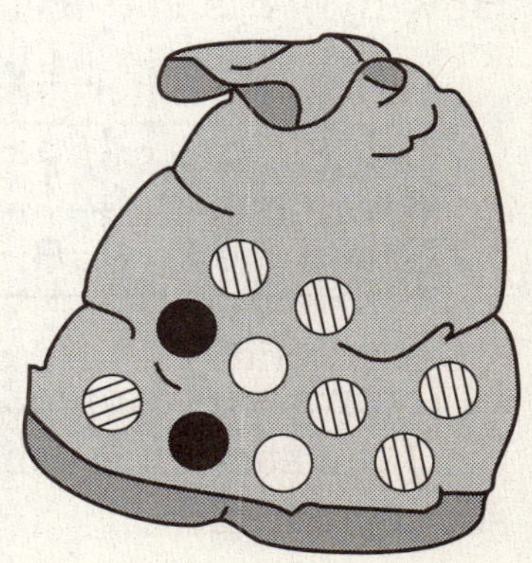

Experiments

You can do probability experiments and use your results to make predictions. Probability experiments will not always turn out like you think they will.

Example

Lynn spun the spinner below 10 times.

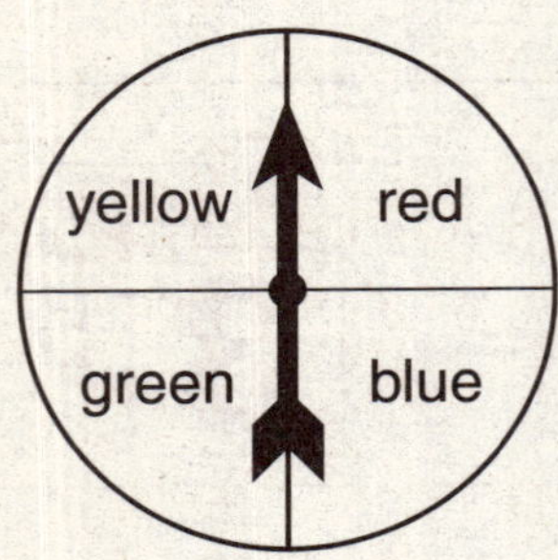

She recorded the results of her experiment in the following table. R stands for red, B stands for blue, G stands for green, and Y stands for yellow.

Lynn's Spinner Experiment

Spin number	1	2	3	4	5	6	7	8	9	10
Outcome	R	B	B	G	Y	Y	B	B	R	Y

In Lynn's experiment, the spinner landed on the red section 2 times, the blue section 4 times, the green section 1 time, and the yellow section 3 times.

Lynn is going to spin the spinner one more time. What color do you predict the spinner will land on?

Each color has one section on the spinner, so they are all equally likely. However, in her experiment, the spinner landed on blue the most times. You could predict that when Lynn spins the spinner again, the arrow will land on blue.

Objectives: D.B.2

Practice

1. Roll a number cube numbered 1 through 6 ten times. Record your results in the table below.

Number Cube Experiment

Roll number	1	2	3	4	5	6	7	8	9	10
Outcome										

If you roll the number cube one more time, what do you predict the outcome will be? Explain why you think so.

____

If you roll the number cube ten more times, how many times do you predict you will roll a 4? Explain why you think so.

____

2. Flip a coin ten times. Record your results in the table below. Use H for heads and T for tails.

Coin Flip Experiment

Coin flip	1	2	3	4	5	6	7	8	9	10
Outcome										

If you flip the coin one more time, what do you predict the outcome will be? Explain why you think so.

____

If you flip the coin ten more times, how many times do you predict it will land heads up? Explain why you think so.

____

Mathematics Practice

1. On which spinner is the arrow **equally likely** to land on a gray or white area?

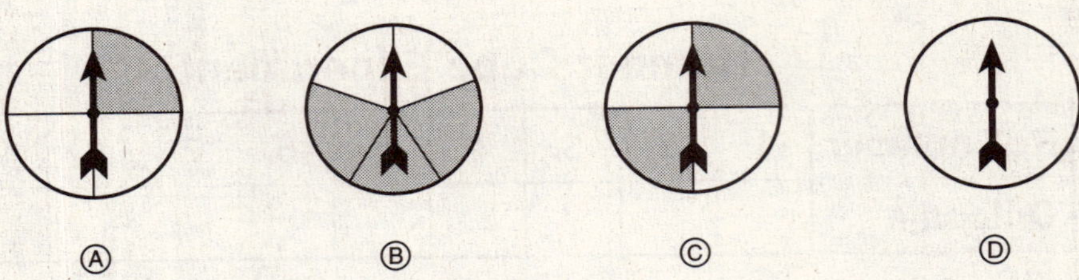

2. Which describes the probability of picking a white cube?

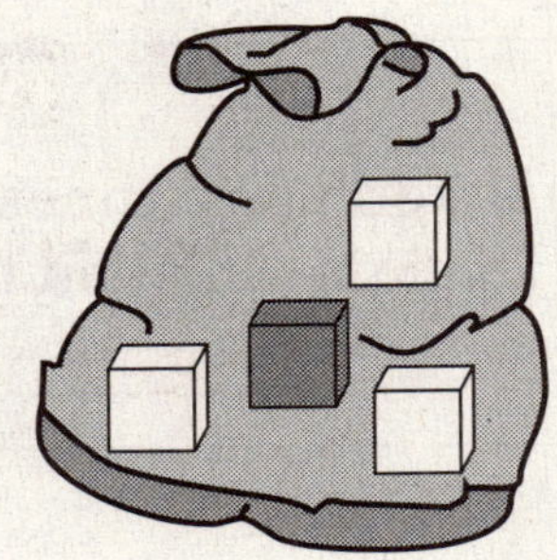

Ⓐ impossible

Ⓑ probable

Ⓒ not probable

Ⓓ certain

3. Adam has 1 penny, 5 nickels, 2 dimes, and 2 quarters in his pocket. If Adam pulls one coin out of his pocket without looking, which coin is he **least likely** to pick?

Ⓐ penny

Ⓑ nickel

Ⓒ dime

Ⓓ quarter

4. There are 20 pencils in a box: 10 are yellow, 6 are blue, and 4 are red. If Mary picks one pencil from the box without looking, which color will she **most likely** pick?

   Ⓐ yellow
   Ⓑ blue
   Ⓒ red
   Ⓓ green

5. Which of the following **best** describes the probability of picking a ◯?

   Ⓐ certain
   Ⓑ most likely
   Ⓒ least likely
   Ⓓ equally likely

6. Mandy is going to a movie sometime this week. Which describes the probability of Mandy going on a day that ends in "day"?

   Ⓐ impossible
   Ⓑ probable
   Ⓒ not probable
   Ⓓ certain

7. How likely is it for a coin to land heads up rather than tails up when it is flipped?

   Ⓐ most likely

   Ⓑ equally likely

   Ⓒ least likely

   Ⓓ certain

8. Jared is spinning a spinner to see how many spaces he can move in a game. On which spinner is it **impossible** for Jared to spin a 6?

   Ⓐ

   Ⓑ

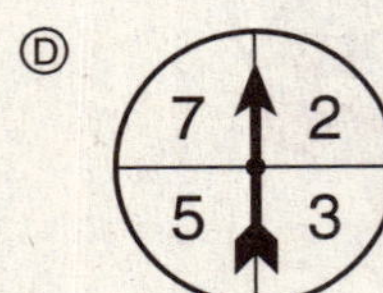

   Ⓒ 6 6 6 6

   Ⓓ 7 2 5 3

9. Shawn is going to roll a number cube numbered 1 through 6. Which describes the probability of Shawn rolling a 3?

   Ⓐ impossible

   Ⓑ probable

   Ⓒ not probable

   Ⓓ certain